METHODS IN MOLECULAR BIOLOGY

For further volumes:
http://www.springer.com/series/7651

For over 35 years, biological scientists have come to rely on the research protocols and methodologies in the critically acclaimed Methods in Molecular Biology series. The series was the first to introduce the step-by-step protocols approach that has become the standard in all biomedical protocol publishing. Each protocol is provided in readily-reproducible step-by step fashion, opening with an introductory overview, a list of the materials and reagents needed to complete the experiment, and followed by a detailed procedure that is supported with a helpful notes section offering tips and tricks of the trade as well as troubleshooting advice. These hallmark features were introduced by series editor Dr. John Walker and constitute the key ingredient in each and every volume of the Methods in Molecular Biology series. Tested and trusted, comprehensive and reliable, all protocols from the series are indexed in PubMed.

Manuscript Instructions https://drive.google.com/file/d/1K4Az5vpAFVzYadJ3jKx2-QSo7Dqa_SNA/view?usp=sharing

Myogenesis and Skeletal Muscle Biology

Methods and Protocols

Edited by

Joe Quadrilatero

Department of Kinesiology and Health Sciences, University of Waterloo, Waterloo, ON, Canada

Humana Press

Editor
Joe Quadrilatero
Department of Kinesiology and Health Sciences
University of Waterloo
Waterloo, ON, Canada

ISSN 1064-3745 ISSN 1940-6029 (electronic)
Methods in Molecular Biology
ISBN 978-1-0716-5395-1 ISBN 978-1-0716-5396-8 (eBook)
https://doi.org/10.1007/978-1-0716-5396-8

This Humana imprint is published by the registered company Springer Science+Business Media, LLC, part of Springer Nature.
The registered company address is: 1 New York Plaza, New York, NY 10004, U.S.A.

Preface

Skeletal muscle is an extraordinary tissue, composed predominantly of postmitotic, multinucleated fibers that play an essential role in human movement and metabolic homeostasis. To fulfill these functions, skeletal muscle must sustain high energetic demands while coordinating complex structural, metabolic, and molecular signaling processes. Skeletal muscle also exhibits a unique capacity for rapid and extensive remodeling, enabling dynamic responses to a wide range of stimuli, including daily physical activity, injury, metabolic stress, and disease. The development, adaptation, and regeneration of skeletal muscle are largely mediated by myogenic stem cells and satellite cells. This regenerative capacity underpins the plasticity of skeletal muscle, supporting not only growth and maintenance but also recovery from injury and adaptation to physiological and environmental challenges. Despite decades of research, many fundamental aspects of myogenesis and skeletal muscle biology remain incompletely understood, reflecting the complexity and heterogeneity of this tissue.

This volume in the *Methods in Molecular Biology* series brings together a comprehensive collection of experimental approaches that are central to the study of myogenesis and skeletal muscle biology. Each chapter follows a consistent and carefully structured format, guiding the reader from fundamental theoretical principles through to a comprehensive list of required reagents, equipment, and materials, ultimately culminating in a clearly written step-by-step methods section. A particular strength of this volume lies in its emphasis on practical insight, with detailed notes sections that highlight critical planning considerations, troubleshooting strategies, and solutions to common technical challenges. These features make this book an indispensable resource for both trainees entering the field and experienced researchers seeking reliable and reproducible methodologies.

This volume offers an extensive and forward-looking toolkit for the study of myogenesis and skeletal muscle biology, spanning stem cell isolation, tissue engineering, advanced imaging, and functional analysis. In particular, these chapters cover state-of-the-art methods for the isolation, expansion, and differentiation of skeletal muscle satellite cells and human-induced pluripotent stem cells, including innovative approaches for generating 3D contractile and vascularized skeletal muscle tissue. A strong emphasis is placed on cutting-edge analytical techniques, with protocols describing ultrastructural fiber and sarcomeric imaging, high resolution fiber type microscopy, flow cytometry-based quantification of differentiation and reactive oxygen species, and advanced microscopy approaches ranging from light-sheet fluorescence imaging to focused ion beam scanning electron microscopy. Additional chapters address functional and mechanistic approaches, including electrical stimulation of human skeletal muscle cells, extracellular vesicle isolation, analysis of mitochondrial morphology, turnover, metabolism, and 3D architecture, as well as protein–protein interaction and molecular dynamics.

Together, these protocols provide researchers with an integrated and practical resource for investigating myogenesis and skeletal muscle biology across multiple model systems and levels of biological organization. This collection of essential and cutting-edge approaches serves as a valuable reference for both basic and translational researchers seeking to understand skeletal muscle development, structure, metabolism, and function.

Ultimately, I hope that this volume will stand as a timely and enduring resource for the skeletal muscle research community, deepening our understanding of the methods used to study myogenesis and skeletal muscle biology. Beyond providing practical guidance, I anticipate that this work will inspire new lines of inquiry and innovative approaches that further reveal insight into skeletal muscle development, adaptation, function, and disease.

I would like to thank all the authors for their valuable contributions and Professor John M. Walker, the series editor, for his advice and guidance in making this book.

Waterloo, ON, Canada ***Joe Quadrilatero***

Contents

Contributors

DANIEL KWAKU ABANKWA • *Cancer Cell Biology and Drug Discovery Group, Department of Life Sciences and Medicine, University of Luxembourg, Esch-sur-Alzette, Luxembourg*
ERIN ALLISON • *Department of Kinesiology and Health Sciences, University of Waterloo, Waterloo, ON, Canada*
SABRINA APEL • *Chemistry Department, University of Massachusetts Lowell, Lowell, MA, USA*
ORI AVINOAM • *Department of Biomolecular Sciences, Weizmann Institute of Science, Rehovot, Israel*
ANNA BENEDETTI • *Department of Anatomical, Histological, Forensic and Orthopedic Sciences, Section of Histology and Embryology, Sapienza University of Rome, Rome, Italy; Translational Oncology Research Unit, IRCSS Regina Elena National Cancer Institute, Rome, Italy*
MARINA BOUCHE • *Department of Anatomical, Histological, Forensic and Orthopedic Sciences, Section of Histology and Embryology, Sapienza University of Rome, Rome, Italy*
ADITYA N. BRAHMBHATT • *School of Kinesiology and Health Science and Muscle Health Research Centre, York University, Toronto, ON, Canada*
NENAD BURSAĆ • *Department of Biomedical Engineering, Duke University, Durham, NC, USA*
FRANCESCA DI CARA • *Department of Microbiology and Immunology, Dalhousie University, Halifax, NS, Canada*
JAIME J. CARVAJAL • *Centro Andaluz de Biología del Desarrollo (CABD), Consejo Superior de Investigaciones Científicas, Universidad Pablo de Olavide, Junta de Andalucía (CSIC-UPO-JA), Sevilla, Spain*
MARIE J CATENACCI • *Sprott Centre for Stem Cell Research, Regenerative Medicine Program, Ottawa Hospital Research Institute, Ottawa, Canada; Departments of Medicine and Cellular and Molecular Medicine, Faculty of Medicine, University of Ottawa, Ottawa, Canada*
YU-SHEN CHENG • *Department of Neural iPSC Research, Institute for Medical Science of Aging, Aichi Medical University, Nagakute, Aichi, Japan*
GUNTARAT CHINVATTANACHOT • *Department of Muscle Development and Regeneration, Institute of Molecular Embryology and Genetics (IMEG), Kumamoto University, Honjo, Chuo-ku, Kumamoto, Japan; Department of Orthopedics, Faculty of Medicine Ramathibodi Hospital, Mahidol University, Ratchathewi, Bangkok, Thailand*
ROHAN CHIPPALKATTI • *Cancer Cell Biology and Drug Discovery Group, Department of Life Sciences and Medicine, University of Luxembourg, Esch-sur-Alzette, Luxembourg*
PAIGE COLE • *Research Institute for Sport and Exercise Sciences, Liverpool John Moores University, Liverpool, UK*
AMBER DETWILER • *Department of Biomedical Engineering, Duke University, Durham, NC, USA*
LORENZA ESPOSITO • *Department of Anatomical, Histological, Forensic and Orthopedic Sciences, Section of Histology and Embryology, Sapienza University of Rome, Rome, Italy*

Eliana O. Fischer • *Faculty of Biomedical Engineering, Technion-Israel Institute of Technology, Haifa, Israel*
Matthew Gage • *Chemistry Department, University of Massachusetts Lowell, Lowell, MA, USA*
Brian Glancy • *National Heart, Lung, and Blood Institute, National Institutes of Health, Bethesda, MD, USA; National Institute of Arthritis and Musculoskeletal and Skin Diseases, National Institutes of Health, Bethesda, MD, USA*
Nicanor González-Morales • *Department of Biology, Life Sciences Centre, Dalhousie University, Halifax, NS, Canada; Department of Microbiology and Immunology, Dalhousie University, Halifax, NS, Canada*
Mackenzie Q. Graham • *Department of Kinesiology and Health Sciences, University of Waterloo, Waterloo, ON, Canada*
Dhanu Gupta • *Institute of Developmental and Regenerative Medicine, Department of Paediatrics., University of Oxford, Oxford, UK; Division of Biomolecular and Cellular Medicine, Department of Laboratory Medicine, Karolinska Institutet, Huddinge, Stockholm, Sweden*
Emily Haughton • *Institute of Developmental and Regenerative Medicine, Department of Paediatrics., University of Oxford, Oxford, UK*
Naoki Horii • *Department of Muscle Development and Regeneration, Institute of Molecular Embryology and Genetics (IMEG), Kumamoto University, Honjo, Chuo-ku, Kumamoto, Japan*
Mayar Ibrabah • *Department of Biology, Life Sciences Centre, Dalhousie University, Halifax, NS, Canada*
Farah Ulfat Jamee • *Department of Neural iPSC Research, Institute for Medical Science of Aging, Aichi Medical University, Nagakute, Aichi, Japan*
Eili Tranheim Kase • *Section for Pharmacology and Pharmaceutical Biosciences, Department of Pharmacy, University of Oslo, Oslo, Norway*
Junki Kawakami • *Department of Orthopedic Surgery, Faculty of Life Sciences, Kumamoto University, Chuo-ku, Kumamoto, Japan*
Alastair Khodabukus • *Department of Biomedical Engineering, Duke University, Durham, NC, USA*
Shulamit Levenberg • *Faculty of Biomedical Engineering, Technion-Israel Institute of Technology, Haifa, Israel*
Biliana Lozanoska-Ochser • *Department of Anatomical, Histological, Forensic and Orthopedic Sciences, Section of Histology and Embryology, Sapienza University of Rome, Rome, Italy; Department of Medicine and Surgery, LUM University, Bari, Italy; Unicamillus-Saint Camillus International University of Health Sciences, Rome, Italy*
Ruhani Makkar • *Institute of Developmental and Regenerative Medicine, Department of Paediatrics., University of Oxford, Oxford, UK*
Tetsuro Masuda • *Department of Orthopedic Surgery, Faculty of Life Sciences, Kumamoto University, Chuo-ku, Kumamoto, Japan*
Takeshi Miyamoto • *Department of Orthopedic Surgery, Faculty of Life Sciences, Kumamoto University, Chuo-ku, Kumamoto, Japan*
Jeffrey Moore • *Department of Biological Sciences, University of Massachusetts Lowell, Lowell, MA, USA*
Tiara Mulder • *Department of Biology, Life Sciences Centre, Dalhousie University, Halifax, NS, Canada*

SERGEY MURSALIMOV • *Department of Biomolecular Sciences, Weizmann Institute of Science, Rehovot, Israel; Institute of Cytology and Genetics, Siberian Branch of Russian Academy of Sciences, Novosibirsk, Russia*
ALEJANDRA NAVARRO-MARTÍNEZ • *Centro Andaluz de Biología del Desarrollo (CABD), Consejo Superior de Investigaciones Científicas, Universidad Pablo de Olavide, Junta de Andalucía (CSIC-UPO-JA), Sevilla, Spain*
NATAŠA NIKOLIĆ • *Section for Pharmacology and Pharmaceutical Biosciences, Department of Pharmacy, University of Oslo, Oslo, Norway*
ANDY NOLAN • *Research Institute for Sport and Exercise Sciences, Liverpool John Moores University, Liverpool, UK*
YOHEI OKADA • *Department of Neural iPSC Research, Institute for Medical Science of Aging, Aichi Medical University, Nagakute, Aichi, Japan*
YUSUKE ONO • *Department of Muscle Development and Regeneration, Institute of Molecular Embryology and Genetics (IMEG), Kumamoto University, Honjo, Chuo-ku, Kumamoto, Japan; Muscle Biology Laboratory, Tokyo Metropolitan Institute for Geriatrics and Gerontology (TMIG), Itabashi-ku, Tokyo, Japan*
GÖKSU ORAL • *Cancer Cell Biology and Drug Discovery Group, Department of Life Sciences and Medicine, University of Luxembourg, Esch-sur-Alzette, Luxembourg*
HAILEY A. PARRY • *National Heart, Lung, and Blood Institute, National Institutes of Health, Bethesda, MD, USA*
CHRISTOPHER G. R. PERRY • *School of Kinesiology and Health Science and Muscle Health Research Centre, York University, Toronto, ON, Canada*
JOE QUADRILATERO • *Department of Kinesiology and Health Sciences, University of Waterloo, Waterloo, ON, Canada*
FASIH A. RAHMAN • *Department of Kinesiology and Health Sciences, University of Waterloo, Waterloo, ON, Canada*
MUHAMMAD IRFANUR RASHID • *Department of Neural iPSC Research, Institute for Medical Science of Aging, Aichi Medical University, Nagakute, Aichi, Japan; Department of Physical Medicine & Rehabilitation, Mayo Clinic, Rochester, MN, USA*
MICHAEL A. RUDNICKI • *Sprott Centre for Stem Cell Research, Regenerative Medicine Program, Ottawa Hospital Research Institute, Ottawa, Canada; Departments of Medicine and Cellular and Molecular Medicine, Faculty of Medicine, University of Ottawa, Ottawa, Canada*
HIDETOSHI SAKURAI • *Department of Clinical Application, Center for iPS Cell Research and Application (CiRA), Kyoto University, Kyoto, Japan*
DARREN W. SEXTON • *School of Pharmacy and Biomolecular Sciences, Liverpool John Moores University, Liverpool, UK*
KAZUNORI SHIMIZU • *Department of Biomolecular Engineering, Graduate School of Engineering, Nagoya University, Nagoya, Aichi, Japan*
SAILAVANYAA SUNDAR • *Department of Biological Sciences, University of Massachusetts Lowell, Lowell, MA, USA*
ANNA TSUKERMAN • *Faculty of Biomedical Engineering, Technion-Israel Institute of Technology, Haifa, Israel; Interdisciplinary Program for Biotechnology Technion-Israel Institute of Technology, Haifa, Israel*
CRISTINA VICENTE-GARCÍA • *Centro Andaluz de Biología del Desarrollo (CABD), Consejo Superior de Investigaciones Científicas, Universidad Pablo de Olavide, Junta de Andalucía (CSIC-UPO-JA), Sevilla, Spain*

MATTHEW J. A. WOOD • *Institute of Developmental and Regenerative Medicine, Department of Paediatrics., University of Oxford, Oxford, UK*
DONGNAN YAN • *Institute of Developmental and Regenerative Medicine, Department of Paediatrics., University of Oxford, Oxford, UK; Nuffield Department of Women's and Reproductive Health, University of Oxford, Oxford, UK*
ILAN ZEMSKI • *Department of Biomolecular Sciences, Weizmann Institute of Science, Rehovot, Israel*

Chapter 1

An Optimized Pre-Plating Method for the Isolation and Culture of Human Primary Satellite Cells

Guntarat Chinvattanachot, Naoki Horii, Tetsuro Masuda, Junki Kawakami, Takeshi Miyamoto and Yusuke Ono

Abstract

Muscle satellite cells play an essential role in postnatal skeletal muscle growth, homeostasis, hypertrophy, and regeneration following injury. Efficient primary culture methods are crucial for the in vitro analysis of satellite cells to investigate the molecular mechanisms underlying the adaptation and regeneration of skeletal muscles. In this protocol, we describe a refined and accessible pre-plating method for the efficient isolation of human satellite cells that does not require specialized equipment or advanced instrumentation. Furthermore, we provide comparative data on various growth medium formulations to identify the optimal and cost-effective conditions for human satellite cell culture. Through systematic optimization of both the isolation technique and culture conditions, we established a practical and reproducible protocol specifically tailored to human skeletal muscles. This methodology enables more accurate and clinically relevant investigations into human muscle biology.

Keywords Human satellite cells, Muscle stem cells, Skeletal muscles, Pre-plating, Primary culture

1 Introduction

Satellite cells are skeletal muscle-resident stem cells that possess the ability to self-renew, proliferate, and differentiate and play a central role in postnatal skeletal muscle growth, homeostasis, hypertrophy, and regeneration following injury [1, 2]. With the rapid increase in the aging population, the prevalence of muscle dysfunction and degeneration has increased drastically, drawing considerable attention to the mechanisms underlying muscle adaptation and regeneration [3]. While murine models, such as mice and rats, have been extensively utilized in skeletal muscle research, species-specific differences in molecular and physiological regulatory pathways have raised concerns regarding their translational relevance to human muscle biology [4]. Consequently, there is growing demand for experimental systems that employ human skeletal muscle-derived cells to accurately model

Joe Quadrilatero (ed.), *Myogenesis and Skeletal Muscle Biology*, Methods in Molecular Biology, vol. 3057, https://doi.org/10.1007/978-1-0716-5396-8_1,

human-specific physiological and pathological responses, particularly when developing therapeutic interventions.

Contrary to the well-established protocols for isolating satellite cells from rodent skeletal muscles [5–13], the isolation and culture of human satellite cells have been developed relatively recently. Protocols for handling human satellite cells remain less standardized and practical experience with them is limited. Although several methodologies for isolating human satellite cells have been reported, many rely on fluorescence-activated cell sorting (FACS) based on surface marker expression [14–16]. While FACS provides high-purity cell populations, the substantial instrumentation and operation costs may pose accessibility barriers for many laboratories. Sorter-induced cellular stress (SICS), a phenomenon associated with reduced cell viability and altered gene expression, raises concerns regarding potential biological artifacts introduced during the sorting process [17–19].

To address these limitations, we previously developed a simple and cost-effective method for isolating mouse and human satellite cells without the use of a cell sorter [20, 21]. This technique was based on the differential adhesion of various cell types to collagen-coated surfaces and allows the enrichment of satellite cell populations by gradually eliminating contaminating fibroblasts and other non-myogenic cells through serial pre-plating steps. However, achieving optimal purity while minimizing cell loss during pre-plating remains a crucial challenge. When we applied this protocol to isolate human satellite cells, we observed that the satellite cells exhibited stronger adhesion to collagen-coated surfaces than expected, resulting in an unintentional loss of target cells during pre-plating. In addition, their proliferation and differentiation rates tended to be significantly slower than those of mouse satellite cells. These differences in activities between mice and humans may be attributed to either the growth medium composition or inherent differences. These findings underscore the need for protocol optimization to account for the unique characteristics of human satellite cells and highlight the broader challenges associated with translating mouse protocols into human systems. Here, we describe a refined and accessible pre-plating method for the efficient isolation of human satellite cells. We also provide comparative data on various growth medium formulations to identify the optimal and cost-effective conditions for human satellite cell culture.

2 Materials

2.1 Isolation of Human Satellite Cells

1. Human skeletal muscle specimens.
2. Adson forceps and jewelers' forceps.
3. Curved-tip scissors.
4. 18-gauge needle and 5 ml syringe.

5. Petri dish, deep-bottom, 50 mm.
6. Collagen type I-coated dish, 60 mm.
7. Six-well plate.
8. Self-standing centrifuge tube, 25 ml.
9. Conical tube, 50 ml.
10. Sterile syringe filter, 0.45 μm.
11. Cell strainer, 40 μm.
12. Dulbecco's phosphate-buffered saline (DPBS).
13. Dulbecco's Modified Eagle Medium (DMEM) with glucose 4.5 g/ml, supplemented with sodium pyruvate.
14. Penicillin–streptomycin (PS).
15. Fetal bovine serum (FBS).
16. Horse serum (HS).
17. Type II collagenase.
18. Trypsin-EDTA (0.25%).
19. Matrigel, Growth Factor Reduced (GFR) Basement Membrane Matrix.
20. 2.5 ng/ml basic fibroblast growth factor (bFGF).

2.2 Culture of Human Satellite Cells

1. Six-well plate.
2. DPBS.
3. Trypsin-EDTA.
4. Growth medium (GM)
   - DMEM.
   - 1% PS.
   - 10% FBS.
   - 10% HS.
   - 2.5 ng/ml of bFGF.
5. Differentiation medium (DM)
   - DMEM.
   - 1% PS.
   - 5% HS.
6. Matrigel Growth Factor Reduced (GFR) Basement Membrane Matrix.

2.3 Immunostaining

1. 4% paraformaldehyde (PFA) in phosphate-buffered saline (PBS).
2. Wash buffer (PBST): 0.025% Tween 20 in PBS.
3. Cell permeabilization buffer: 0.3% Triton X-100 in PBS.

4. Blocking reagent: Normal Goat Serum.
5. Primary antibodies
   - Anti-MYOD, rabbit polyclonal (Cat. No.: PA5-23078, Invitrogen, 1:3000).
   - Anti-PAX7, mouse monoclonal (Cat. No.: sc-81648, Santa Cruz, 1:200).
6. Secondary antibodies
   - Alexa Fluor 488-conjugated anti-mouse IgG, goat polyclonal (1:1000).
   - Alexa Fluor 555-conjugated anti-rabbit IgG, goat polyclonal (1:1000).
7. Click-iT™ EdU.
8. Mounting Medium: Fluoro-KEEPER Antifade Reagent, Non-Hardening Type with DAPI.
9. Fluorescence microscope.

3 Methods

3.1 Isolation of Satellite Cells from Human Muscle Specimens

1. Following surgical retrieval, the specimens are stored in DPBS and promptly transported on ice to the laboratory for further processing.
2. Type II collagenase is dissolved in DMEM containing 1% PS at a final concentration of 0.2%. The solution is then filtered through a 0.45 μm sterile syringe filter and stored in an incubator for prewarm until use.
3. Skeletal muscle tissue is thoroughly rinsed three times with DMEM containing 1% PS to remove any potential contaminants. The tissue is meticulously dissected using forceps and scissors to remove non-muscular structures such as tendons and fat (Fig. 1a). The remaining muscle is finely minced using scissors (Fig. 1b) (*see* **Note 1**).
4. Minced skeletal muscle tissue is transferred to 25 ml tubes containing 0.2% Type II collagenase in DMEM, added at a ratio of 10 ml of solution per gram of muscle (1–1.5 g of skeletal muscle per tube). The mixture is then incubated at 37°C on a shaker for 45 minutes (Fig. 1c).
5. After the initial incubation, the mixture is aspirated and expelled several times using a syringe fitted with an 18 G needle to further dissociate the tissue (*see* **Note 2**). The mixture is then placed on a shaker for another 15 minutes under the same incubation conditions.

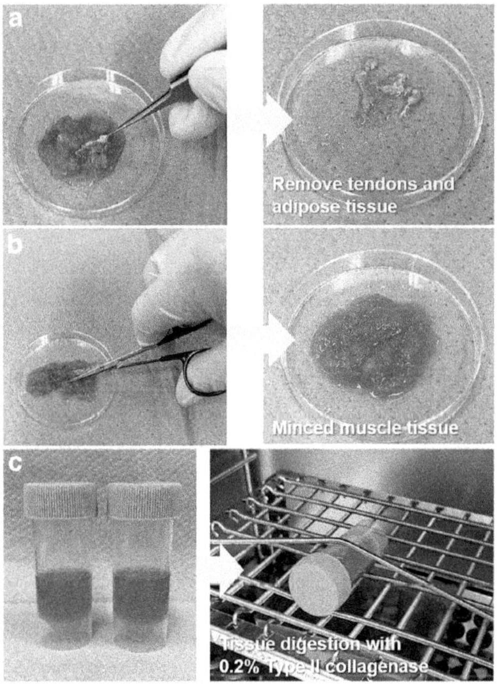

Fig. 1 Dissection of human skeletal muscle tissue. (**a**) Muscle tissues are carefully dissected to remove non-muscle components, such as tendons and adipose tissue, in order to minimize contamination by non-muscle cell populations. (**b**) The isolated muscle tissue is finely minced using sterile scissors. (**c**) The minced muscle tissues are transferred into 25 ml tubes containing 0.2% Type II collagenase in DMEM and incubated at 37°C on a shaker

6. The enzymatic reaction is terminated by diluting the mixture with DPBS to a final volume of 25 ml per tube.
7. The final mixture is transferred into new 50 ml conical tubes through a 40 μm cell strainer to remove residual debris. The filtered mixture is centrifuged at 700 × *g* for 10 minutes.
8. Following the first centrifugation, the supernatant is carefully aspirated and replaced with 30 ml of 2% FBS in DPBS in each tube. The tubes are subsequently centrifuged at 700 × *g* for 10 minutes. During this period, 3–4 ml of GM should be prepared in collagen type I-coated dishes and prewarmed in an incubator. The number of dishes prepared should correspond to the number of tubes in which the sample was allocated.
9. Discard the supernatant and gently resuspend the pellet with 2 ml of GM per tube. The suspension is then transferred to the prepared collagen-coated dishes and incubated for 16 hours or overnight in a 37°C humidified incubator with 5% CO_2, which will be the standard incubation setting throughout the process.

10. On the next day, unattached cells are resuspended by gentle pipetting over the entire surface of the dish. Transfer the media to the new collagen-coated dishes and incubate for 3 hours.
11. After the 3-hour incubation, the medium is transferred to Matrigel-coated six-well plates (*see* **Note 3**). Typically, the contents of one collagen-coated dish are distributed across two to three wells of the six-well plates. The plate is incubated for 24 hours.
12. On the third day, GM containing floating debris is carefully aspirated and replaced with fresh GM. The plate is then incubated for 24 hours.
13. After incubation, the medium is removed, and the plate is gently washed with DPBS. The cells are then trypsinized with 200 μl of Trypsin-EDTA according to standard protocols. Once the cells have rounded and detached, the reaction is stopped by adding 2 ml of 10% FBS in DMEM. The cell suspension is then centrifuged at 1800 rpm for 5 minutes. After centrifugation, the supernatant is discarded, and the cell pellet is resuspended in GM before being transferred to a new collagen-coated dish for the next step.
14. The collagen-coated dish is placed in an incubator for 1 hour. During this time, periodic manual shaking at 5-minute intervals is performed to keep the satellite cells in suspension while allowing faster attachment of non-myogenic cells, such as fibroblasts, to the collagen-coated surface (Fig. 2) (*see* **Note 4**).
15. After the 1-h pre-plating period, the dish is thoroughly pipetted to collect all non-attached cells. This pipetting using a 1000 μl pipette is gentle enough to allow the collection of loosely attached muscle cells while avoiding excessive loss of satellite cell populations and preventing excessive detachment of non-muscle cells. The medium containing the floating cells is counted and allocated to new Matrigel-coated plates. If a significant number of cells remain attached to the collagen-coated dish, it can be washed with fresh media, and the

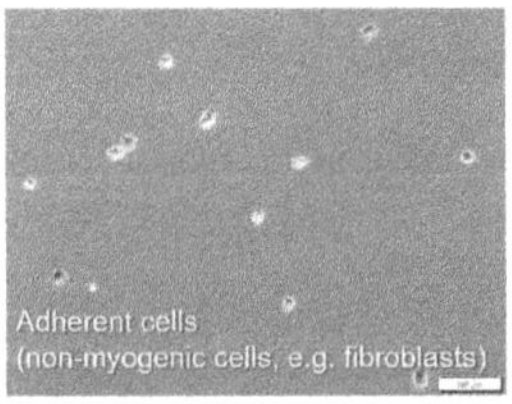

Fig. 2 Non-myogenic cells. A bright-field image illustrates the remaining cells that attached to the collagen-coated dish during the pre-plating step. The majority of these adherent cells are fibroblasts, which are selectively excluded at this stage. Scale bar, 100 μm

remaining cells can be collected and added to the previously collected suspension.

16. The satellite cells on the Matrigel-coated plate are incubated for further expansion (Fig. 3a, b).

3.2 Culture of Primary Human Satellite Cells for Population Expansion and Myogenic Differentiation

1. During early passages, primary human satellite cells are seeded at a density of 1000 cells/cm^2 and incubated under previously described culture conditions.
2. The culture medium is replaced every 48 hours until the cells reach approximately 70% confluency, at which point they are harvested by trypsinization and subcultured. Cells are either reseeded at a density of 1000 cells/cm^2 or passaged at a 1:4 dilution ratio if cell counting is not feasible.
3. For differentiation assays, the cells are cultured until they reach full confluency. Prior to switching to DM, the plates are washed twice with DPBS to remove the residual GM. Myogenic differentiation is induced under the same incubation conditions as proliferation, with 50% of the DM replaced daily (Fig. 4).

3.3 Immunostaining to Assess the Purity of Human Satellite Cells

The purity of the modified pre-plating technique is validated by immunostaining myogenic cells for PAX7 and MYOD markers (Fig. 5a, b). Optionally, EdU incorporation is used to assess cell proliferation. Immunostaining protocol is as follows:

1. The cells cultured on the plates are fixed with 4% PFA in PBS for 15 minutes. Following fixation, PFA is removed, and the cells are washed three times with PBS for 5 minutes to ensure complete removal of PFA.
2. To prepare the blocking/permeabilizing solution, goat serum is added to 0.3% Triton X-100 in PBS at a concentration of 5% (w/v). The cells are incubated with a blocking/permeabilizing solution for 20 minutes.
3. The primary antibody solution is prepared in PBST containing Anti-MYOD and Anti-PAX7 antibodies (*see* **Note 5**). Following removal of the blocking/permeabilizing solution, the plate is washed once with PBST and incubated with primary antibodies on a shaker at 4°C overnight.
4. The following day, the primary antibodies are removed, and the plate is washed three times with PBST for 5 minutes each.
5. The secondary antibody solution contains Alexa Fluor 488-conjugated anti-mouse IgG and Alexa Fluor 555-conjugated anti-rabbit IgG, both diluted to 1:1000 in PBST (*see* **Note 5**). The cells are incubated with the secondary antibody solution on a shaker at room temperature for 1 hour (*see* **Note 6**). After incubation, the secondary antibodies are removed, and

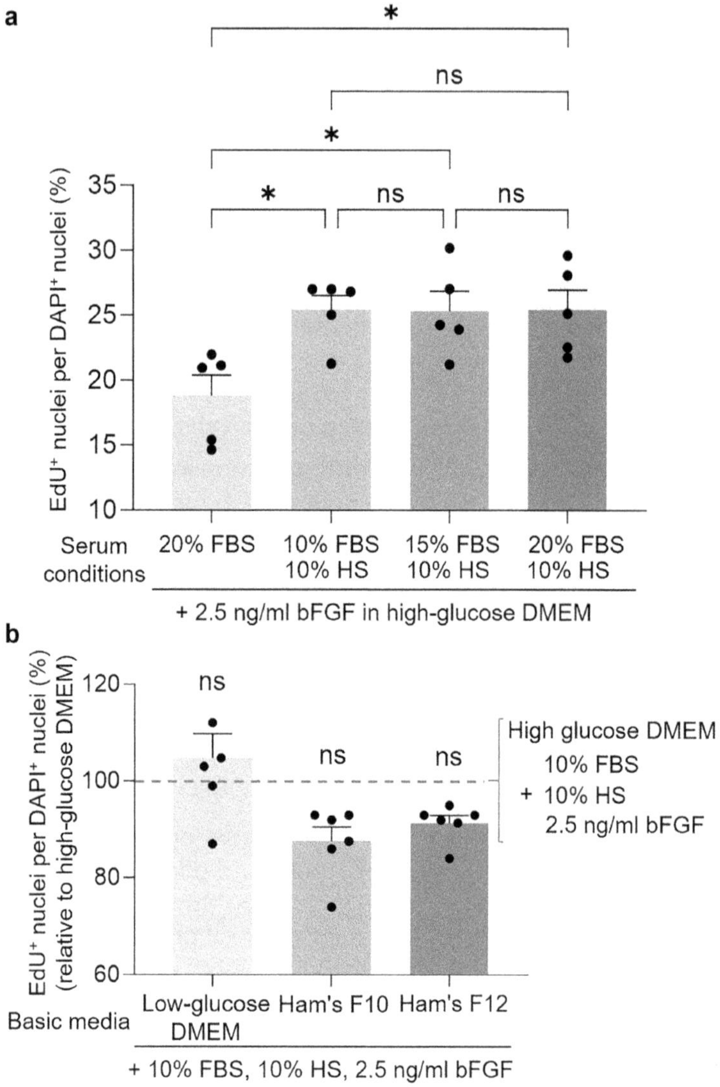

Fig. 3 Proliferation rates of human satellite cells cultured in different growth media. (**a**) Cell proliferation is evaluated using EdU incorporation assays. Results show that satellite cells cultured in high-glucose DMEM (4.5 g/l glucose) supplemented with both FBS and HS exhibit significantly higher proliferation rates than those cultured with FBS alone. No significant differences are observed among the groups supplemented with different ratios of FBS and HS. $N = 5$; $^*P < 0.05$. (**b**) Using high-glucose DMEM supplemented with 10% FBS and 10% HS as a reference condition (=100%), other basal media, including low-glucose DMEM (1 g/l glucose), high-glucose DMEM, Ham's F10, and Ham's F12, are compared. No statistically significant differences in proliferation are observed among the different basal media, suggesting that the composition of the basal medium had a minimal effect on proliferation under these conditions. ns, not significant

Fig. 4 Myogenesis of human satellite cells. Myogenic differentiation of human satellite cells was induced in DM for 3 days. Myosin heavy chain (MyHC)+ multinucleated myotubes were observed. Scale bar, 200 μm

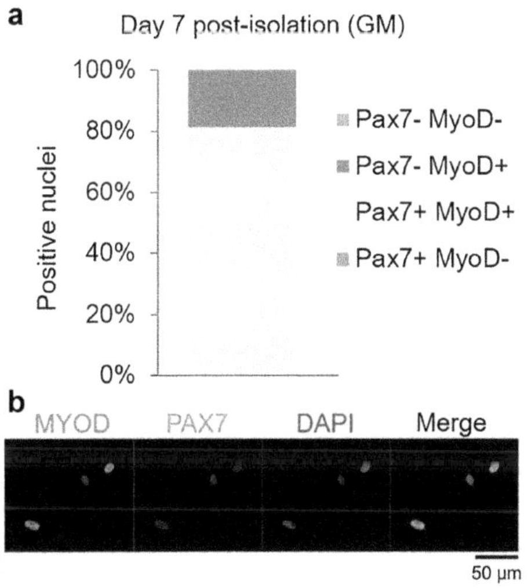

Fig. 5 Validation of satellite cell purity. (**a**) Purity of isolated human satellite cells was assessed following isolation using the modified pre-plating method. Cells were fixed and immunostained for PAX7 and MYOD. Satellite cell purity was defined as the percentage of PAX7$^+$ or MYOD$^+$ cells relative to the total number of DAPI$^+$ nuclei. The majority of the cell population consisted of either PAX7$^+$/MYOD$^+$ or PAX7$^-$/MYOD$^+$ cells, indicating a high purity of approximately 99% in GM at day 7 post-isolation. (**b**) Representative immunofluorescence images of cells positive for PAX7 or MYOD. Scale bar, 50 μm

the plate is washed three times following primary antibody incubation.

6. If EdU staining is required, it is performed at this stage using the Click-iT™ EdU kit (*see* **Note 7**). The Click-iT™ reaction cocktail is prepared according to the manufacturer's protocol and added to each well, followed by incubation at room temperature, protected from light, for 30 minutes. After incubation, the reaction cocktail is removed, and the plate is washed three times for 5 minutes.

7. Finally, the Mounting Medium containing DAPI is added to each well. The cells may be immediately analyzed or stored at 4°C for further examination.

4 Notes

1. Do not overmince the skeletal muscles to avoid damaging the satellite cell population; no more than 100 cuts with scissors per 4–6 g of specimen.
2. If the solution did not pass through an 18-gauge syringe, mince the large pieces of skeletal muscle tissue with scissors.
3. For coating, prepare the Matrigel solution by mixing the Matrigel Matrix with DMEM on ice to a final concentration of 1 mg/ml.
4. The interval of manually applied periodic shaking is critically important for minimizing nonmyogenic cell contamination while preserving satellite cell numbers.
5. If nonspecific background staining is high, 1% goat serum should be added to the solution.
6. To prevent fluorescence bleaching, the plate should be covered with aluminum foil to minimize light exposure during immunostaining.
7. Cells are labeled with 10 μM EdU in GM for 3 hours in a 37°C humidified incubator with 5% CO_2 prior to fixation.

Acknowledgments

This study was supported by the FOREST Program of the Japan Science and Technology Agency (JST, grant no. JPMJFR205C), the Japan Agency for Medical Research and Development (AMED, Grant No. 23ek0210182, 25ek0109823, and 25gm2110003), and Grants-in-Aid for Scientific Research KAKENHI (Grant No. 22K18414 and 25H01102).

References

1. Relaix F, Zammit PS (2012) Satellite cells are essential for skeletal muscle regeneration: the cell on the edge returns centre stage. Development 139(16):2845–2856
2. Ono Y (2014) Satellite cell heterogeneity and hierarchy in skeletal muscle. J Phys Fitness Sports Med 3(2):229–234
3. Chinvattanachot G, Rivas D, Duque G (2024) Mechanisms of muscle cells alterations and regeneration decline during aging. Ageing Res Rev 102:102589
4. Borsch A, Ham DJ, Mittal N, Tintignac LA, Migliavacca E, Feige JN et al. (2021) Molecular and phenotypic analysis of rodent models

reveals conserved and species-specific modulators of human sarcopenia. Commun Biol 4(1):194
5. Garcia P, Mercier O, Ravent J, Le Grand F (2024) Protocol for cell proliferation and cell death analysis of primary muscle stem cell culture using flow cytometry. STAR Protoc 5(4):103411
6. Barajaa MA, Otsuka T, Laurencin CT (2025) A novel protocol for the direct isolation of a highly pure and regenerative population of satellite stem cells. Proc Natl Acad Sci U S A 122(25):e2426081122
7. Boscolo Sesillo F, Wong M, Cortez A, Alperin M (2020) Isolation of muscle stem cells from rat skeletal muscles. Stem Cell Res 43:101684
8. Elizalde G, Munoz AZ, Almada AE (2023) Protocol for the isolation of mouse muscle stem cells using fluorescence-activated cell sorting. STAR Protoc 4(4):102656
9. Kitajima Y, Ogawa S, Ono Y (2016) Visualizing the functional heterogeneity of muscle stem cells. Methods Mol Biol 1516:183–193
10. Kitajima Y, Ono Y (2018) Visualization of PAX7 protein dynamics in muscle satellite cells in a YFP knock-in-mouse line. Skelet Muscle 8(1):26
11. Ganassi M, Zammit PS, Hughes SM (2021) Isolation of myofibres and culture of muscle stem cells from adult zebrafish. Bio Protoc 11(17):e4149
12. Liu L, Cheung TH, Charville GW, Rando TA (2015) Isolation of skeletal muscle stem cells by fluorescence-activated cell sorting. Nat Protoc 10(10):1612–1624
13. Bosnakovski D, Xu Z, Li W, Thet S, Cleaver O, Perlingeiro RC et al. (2008) Prospective isolation of skeletal muscle stem cells with a Pax7 reporter. Stem Cells 26(12):3194–3204
14. Uezumi A, Nakatani M, Ikemoto-Uezumi M, Yamamoto N, Morita M, Yamaguchi A et al. (2016) Cell-surface protein profiling identifies distinctive markers of progenitor cells in human skeletal muscle. Stem Cell Rep 7(2):263–278
15. Garcia SM, Tamaki S, Lee S, Wong A, Jose A, Dreux J et al. (2018) High-yield purification, preservation, and serial transplantation of human satellite cells. Stem Cell Rep 10(3):1160–1174
16. Alexander MS, Rozkalne A, Colletta A, Spinazzola JM, Johnson S, Rahimov F et al. (2016) CD82 Is a marker for prospective isolation of human muscle satellite cells and is linked to muscular dystrophies. Cell Stem Cell 19(6):800–807
17. Andra I, Ulrich H, Durr S, Soll D, Henkel L, Angerpointner C et al. (2020) An evaluation of T-cell functionality after flow cytometry sorting revealed p38 MAPK activation. Cytometry A 97(2):171–183
18. Pfister G, Toor SM, Sasidharan Nair V, Elkord E (2020) An evaluation of sorter induced cell stress (SICS) on peripheral blood mononuclear cells (PBMCs) after different sort conditions—are your sorted cells getting SICS?. J Immunol Methods 487:112902
19. Box A, DeLay M, Tighe S, Chittur SV, Bergeron A, Cochran M et al. (2020) Evaluating the effects of cell sorting on gene expression. J Biomol Tech 31(3):100–111
20. Yoshioka K, Kitajima Y, Okazaki N, Chiba K, Yonekura A, Ono Y (2020) A modified pre-plating method for high-yield and high-purity muscle stem cell isolation from human/mouse skeletal muscle tissues. Front Cell Dev Biol 8:793
21. Yoshioka K, Nagahisa H, Miura F, Araki H, Kamei Y, Kitajima Y et al (2021) Hoxa10 mediates positional memory to govern stem cell function in adult skeletal muscle. Sci Adv 7(24):eabd7924

Chapter 2

Isolation and Expansion of Muscle Satellite Cells Using the Ice-Cold Treatment Method

Anna Benedetti, Lorenza Esposito, Marina Bouche and Biliana Lozanoska-Ochser

Abstract

Satellite cells (SCs) are skeletal muscle stem cells responsible for muscle regeneration after injury. In this chapter, we describe the ice-cold treatment (ICT) method, a procedure developed for isolating and expanding highly pure populations of mouse and human SCs in vitro. The ICT method involves placing a dish containing a mixed population of adherent muscle mononuclear cells on ice for 15–30 min. This selectively causes only the SCs to detach, resulting in cell cultures with purity levels ranging from 95% to 100%. Furthermore, this method supports serial passaging, allowing prolonged expansion of SCs without negatively impacting their ability to proliferate or differentiate.

Keywords Satellite cells isolation, Satellite cells in vitro expansion, Skeletal muscle, Pax7, MyoD

1 Introduction

Skeletal muscle regeneration largely depends on a resident stem cell population known as satellite cells (SCs) [1–3]. Investigating their functional potential requires effective methods for isolating and expanding highly pure SC populations that retain their myogenic properties through serial in vitro passages [4–6].

Currently, three primary techniques are commonly employed to isolate SCs: the pre-plating method, fluorescence-activated cell sorting (FACS), and magnetic-activated cell sorting (MACS).

The pre-plating technique relies on the differential adhesive properties of skeletal muscle cells, where SCs exhibit the lowest adhesion [4–7]. Although this method is cost-effective and straightforward, its main drawbacks include being time-intensive and yielding cultures of inconsistent purity, often contaminated and rapidly overgrown by fibroblasts by day 7 [5].

FACS sorts muscle mononuclear cells pre-labeled with antibodies specific to SC markers [8–13]. Currently, FACS represents the

Joe Quadrilatero (ed.), *Myogenesis and Skeletal Muscle Biology*, Methods in Molecular Biology, vol. 3057, https://doi.org/10.1007/978-1-0716-5396-8_2,

gold standard in SC isolation. However, disadvantages include high costs, the necessity for specialized sorting equipment, significant time investment, and requisite technical expertise. Additionally, cell labeling and sorting may stress or damage the cells, potentially reducing their viability or altering their functional properties [6].

The third approach, magnetic-activated cell sorting (MACS), utilizes magnetic columns coupled with magnetic bead kits targeting SC markers [14]. This method presumes the complete removal of non-targeted cells from preparations, rendering it less precise than FACS. MACS is also costly, time-consuming, and may stress the cells, resulting in variable purity and frequent fibroblast contamination by the seventh day of culture [5, 6].

In this chapter, we introduce the ice-cold treatment method (ICT), a straightforward, cost-effective, and efficient procedure for isolating and long-term expansion of highly pure mouse and human SC populations with preserved myogenic capacity [15]. A key advantage of the ICT method is its additional utility as a gentle passaging method, enabling prolonged serial expansion of SCs ex vivo without negatively affecting their proliferation or differentiation potential. Consequently, the use of ICT method substantially reduces the number of mice or skeletal muscle biopsies needed to obtain adequate SC cell numbers [16, 17]. This approach significantly contrasts with traditional passaging reagents such as trypsin, known to accelerate SC differentiation after only two passages [4, 15–18].

2 Materials

1. Microsurgery scissors.
2. Microsurgery tweezers.
3. Scalpel and surgical blades.
4. Pipettes (Gilson, P10, 20, 200, 1000).
5. 70 μm cell strainers.
6. 40 μm cell strainers.
7. 100 mm tissue culture dishes.
8. 35 mm tissue culture dishes.
9. 60 mm tissue culture dishes.
10. 50 ml polypropylene centrifuge tubes.
11. 10 ml serological pipettes.
12. Dissection boards (Styrofoam board).
13. Cover slips 24 × 50.
14. Pipette tips.
15. Parafilm.

16. WT C57BL/6J mice aged 4–6 weeks (Jackson Laboratory).
17. Dulbecco's modified Eagle's medium (DMEM + GlutaMAX, + 4.5 g/l D-glucose, − pyruvate).
18. Bovine serum albumin (BSA).
19. DMEM 10% FBS.
20. Hydrophobic PAP pen for immunostaining.
21. Human skeletal muscle biopsies (In the original *Skeletal Muscle* article (Benedetti et al. [16, 17]), gluteus maximus biopsies were obtained from patients undergoing surgery).
22. Vectashield mounting medium.
23. Hoechst 33342 staining dye.
24. Primary antibodies:
    - Pax7 (Developmental Studies Hybridoma Bank).
    - MyoD (Santa Cruz Biotechnology, catalog number: sc-760).
    - MyoG (F5D) (Developmental Studies Hybridoma Bank).
    - MyHC (MF20) (Developmental Studies Hybridoma Bank).
25. Fluorescently labeled secondary antibodies:
    - Goat anti-rabbit Alexa Fluor 488 (1:1000, Abcam).
    - Goat anti-mouse Alexa Fluor 555 (1:1000, Thermo Fisher Scientific).
26. 0.1% gelatin.
27. L-glutamine.
28. Chicken embryo extract (Seralab).
29. Goat serum.
30. Horse serum.
31. Fetal bovine serum, heat inactivated at 56°C for 30 min.
32. Dulbecco's phosphate-buffered saline (PBS) with $MgCl_2$ and $CaCl_2$.
33. PBS, modified without $MgCl_2$ and $CaCl_2$.
34. Penicillin–streptomycin (Pen/Strep) solution.
35. Gentamycin solution.
36. Formaldehyde solution 4% (PFA).
37. Methanol.
38. Collagenase type II.
39. Collagenase/dispase.
40. 70% ethanol (70 ml of 100% ethanol to 30 ml of ddH_2O).

41. Growth medium (DMEM supplemented with 20% horse serum, 3% chicken embryo extract, 1% penicillin–streptomycin, and 1% L-glutamine).
42. Differentiation medium (DMEM supplemented with 5% horse serum, 1% chicken embryo extract, 1% penicillin–streptomycin, and 1% L-glutamine).
43. Enzyme neutralization buffer (DMEM medium with 10% FBS and 1% penicillin–streptomycin).
44. Hoechst solution (Hoechst stain 1:1000 in PBS).

2.1 Equipment

1. Humidified chamber (prepared by wetting paper towels with distilled water and placing them in a plastic container with a lid).
2. Centrifuge.
3. Vertical autoclave.
4. Temperature-regulated shaking water bath.
5. Biosafety cabinet.
6. Laboratory chemical fume hood.
7. CO_2 incubator.
8. Zeiss Axioskop 2 Plus microscope.
9. Phase-contrast microscope.
10. Pipette controllers for serological pipettes.
11. Hemocytometer (Neubauer improved counting chamber).
12. Ice machine.

2.2 Software

1. ZEISS ZEN 2 Blue edition (Carl Zeiss) (Link for download: https://www.zeiss.com/microscopy/int/products/microscope-software.html).
2. ImageJ 1.53a (Link for download: https://imagej.nih.gov/ij/download.html).

3 Methods

3.1 Isolation and Culture of Mouse Satellite Cells

3.1.1 Preparation Prior to Muscle Dissection

1. Prepare 70% ethanol (70 ml of 100% ethanol to 30 ml of ddH_2O).
2. Autoclave dissection instruments (scissors and tweezers).

3. Add 3 ml of DMEM supplemented with 1% Pen/Strep into 60 mm cell culture dishes (one dish/mouse; or one dish per gram of human skeletal muscle biopsy). These will be used to collect the dissected muscle.
4. Coat cell culture dishes with 0.1% gelatin: Coat 35 and 100 mm cell culture dishes with 1 and 5 ml of 0.1% gelatin respectively, making sure that the entire surface of the dish is evenly coated, and incubate for 30 min at 37°C, 5% CO_2 in a cell incubator. Next, remove excess gelatin (without washing the dishes) and leave the dishes open to air dry for 10–15 min under the hood. These dishes will be used for cell culture (*see* **Note 1**).
5. Prepare collagenase type II digestion solution (dissolve collagenase type II powder in PBS (with $MgCl_2$ and $CaCl_2$) at a concentration of 0.4 mg/ml, for a total of 10 ml/g of tissue).
6. Prepare collagenase/dispase digestion solution (dissolve collagenase/dispase powder in PBS (without $MgCl_2$ and $CaCl_2$) at a concentration of 1 mg/ml, for a total of 10 ml/g of tissue).
7. Prepare enzyme neutralization buffer (supplement DMEM medium with 10% FBS and 1% penicillin–streptomycin).
8. Prepare cell growth medium (supplement DMEM medium with 20% horse serum, 3% chicken embryo extract, 1% penicillin–streptomycin, and 1% L-glutamine. Store at 4°C).
9. Sterilize scalpel and surgical blades.
10. Pre-warm a water bath at 37°C.

3.1.2 Day 1: Dissection of Mouse Hindlimb Muscles

1. Euthanize the mice (in the original article (Benedetti et al. [16, 17]), the mice were euthanized by cervical dislocation).
2. Pin the mouse on a dissection board and spray it with 70% ethanol.
3. Using dissection scissors, make an incision of the skin above the ankle and peel off the skin over the entire length of the hindlimbs to expose the muscles below.
4. Collect the tibialis anterior, quadriceps, and gastrocnemius muscle from both legs (see Fig. 1).
5. Place the dissected muscles in a 60 mm dish containing DMEM supplemented with 1% Pen/Strep and remove any hair to minimize risk of contamination.
6. Transfer the dishes under a biosafety cabinet and finely cut the muscles into 1 mm pieces using a sterile scalpel (Fig. 2b). Remove and discard any fat from the dissected muscle.
7. Transfer the minced muscles into a 50 ml tube and centrifuge at 300 × *g* for 5 min at room temperature (RT).

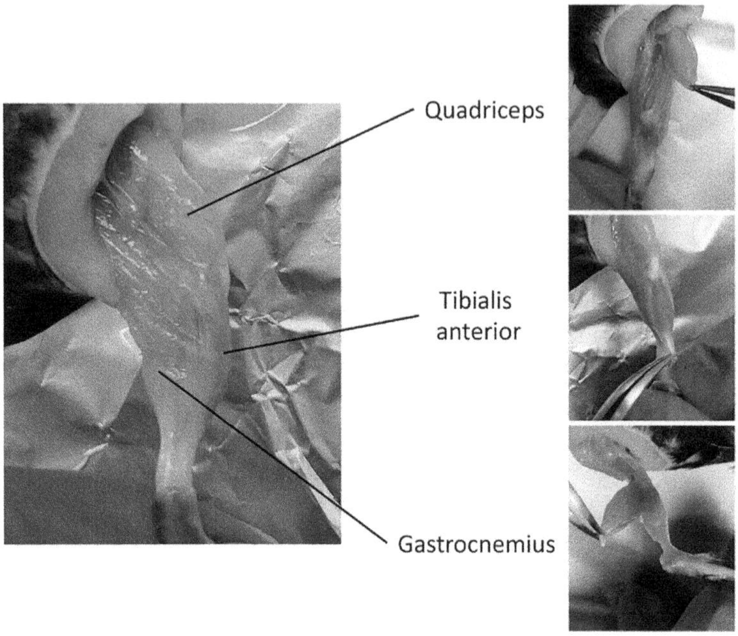

Fig. 1 Dissection of mouse quadriceps, tibialis anterior, and gastrocnemius muscles. Photographs depicting the dissection of quadriceps, tibialis anterior, and gastrocnemius muscles of the hindlimb. (Reprinted with permission from [17])

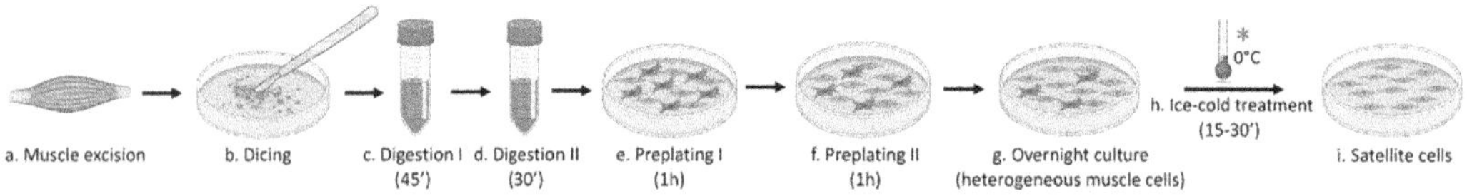

Fig. 2 Satellite cell isolation using the ice-cold treatment method. Schematic diagram of the steps (a-i) involved in the protocol for the isolation of satellite cells using the ice-cold treatment method. (Reprinted from [17] with permission)

3.1.3 Day 1: Muscle Digestion I (Fig. 2c)

1. Aspirate the supernatant with a 10 ml serological pipette and resuspend the minced muscles in collagenase type II digestion solution at a final volume of 10 ml/g of tissue (*see* **Note 2**).
2. Seal the tubes with parafilm and transfer into a shaking (100 rpm) pre-warmed (37°C) water bath for 45 min (*see* **Note 3**).
3. Spray the tubes with 70% ethanol and transfer them under a biosafety cabinet.
4. Block the digestion enzyme by adding 10 ml of enzyme neutralization buffer per gram of tissue.
5. Pipette up and down with a 10 ml serological pipette five times.
6. Centrifuge at 300 × *g* for 5 min at RT.

3.1.4 Day 1: Muscle Digestion II (See Fig. 2d)

1. Aspirate the supernatant with a 10 ml serological pipette and resuspend the pellet in collagenase/dispase digestion solution at a final volume of 10 ml/g of tissue.
2. Seal the tubes with parafilm and transfer into a shaking (100 rpm), pre-warmed (37°C) water bath for 30 min.
3. Spray the tubes with 70% ethanol and transfer them under a biosafety cabinet.
4. Block the digestion by adding 10 ml of enzyme neutralization buffer per gram of tissue.
5. Pipette up and down with a 10 ml serological pipette five times.
6. Pass the digested muscles through a 70 μm cell strainer into a new 50 ml tube.
7. Pass the digested muscles through a 40 μm cell strainer into a new 50 ml tube (*see* **Note 4**).
8. Centrifuge at 300 × *g* for 5 min at RT.
9. Discard the supernatant and resuspend the pellet in 10 ml of DMEM with 10% FBS and count the number of muscle mononuclear cells using trypan blue (*see* **Note 5**).

3.1.5 Day 1: Pre-Plating Step (See Fig. 2e, f)

1. Plate the cells at 10^5/ml into a 100 mm dish (uncoated) in 20 ml DMEM with 10% FBS and 1% Pen/Strep and incubate at 37°C for 1 h.
2. Collect the non-adhered cells into a 50 ml tube.
3. Centrifuge at 300 × *g* for 5 min at RT.
4. Discard the supernatant and resuspend the cell pellet with DMEM supplemented with 10% FBS and 1% Pen/Strep.
5. Count the resuspended cells and plate them at 10^5 cells/ml into a 100 mm dish (uncoated) in 20 ml DMEM with 10% FBS and 1% Pen/Strep and incubate at 37°C for 1 h.
6. Collect and transfer non-adhered cells to a 50 ml tube (*see* **Note 6**).
7. Centrifuge at 300 × *g* for 5 min at RT.
8. Discard the supernatant and resuspend the cell pellet in growth medium.
9. Plate the cells into 0.1% gelatin-coated 100 mm dishes at 10^6 cells/dish and grow them overnight (Fig. 2g).

3.1.6 Day 2: Ice-Cold Treatment (ICT) (See Fig. 2h)

1. Wash the dishes containing the heterogeneous mix of adhered muscle cells three times with 10 ml of PBS at room temperature (with $MgCl_2$ and $CaCl_2$), by gently pipetting PBS all around the dish.
2. Add 10 ml of ice-cold PBS (without $MgCl_2$ and $CaCl_2$) to each 100 mm dish (*see* **Note 7**).

3. Fill a polystyrene box with ice (0°C) and place the dishes on top of the ice for 15–30 min with occasional gentle manual shaking of the dish (swirling motion) (*see* **Notes 8 and 9**).
4. Collect the detached cells into a 50 ml tube.
5. Wash the original dish once more with 10 ml of PBS to collect the remaining detached cells, by gently pipetting PBS all around the dish.
6. Centrifuge the cells at 300 × *g* for 5 min at RT.
7. Discard the supernatant and resuspend the cells in growth medium.
8. Count the cells (*see* **Note 10**).
9. Plate the cells into 35 mm dishes precoated with 0.1% gelatin at a density of 10^3 cells/dish and place them into a cell incubator (*see* **Note 11**).

3.1.7 Cell Culture

1. Culture the cells in growth medium, changing the medium every 2 days.
2. To induce myoblast differentiation, remove growth medium and replace it with differentiation medium once the cells reach 80% confluency.

3.1.8 Cell Passage

1. Place the dishes containing proliferating SCs on ice for up to 30 min.
2. Follow the steps described in Sect. 3.1.6.

3.2 Isolation and Culture of Human Skeletal Muscle Satellite Cells

Follow the same procedure as for mouse skeletal muscle (*see* **Notes 12–16**).

3.3 Analysis of Mouse and Human Satellite Cell-Specific Marker Expression by Immunofluorescence Staining

1. Grow the cells in a 35 mm dish.
2. Aspirate the medium and wash the dishes once with 1 ml of PBS with $MgCl_2$ and $CaCl_2$.
3. Fix the cells by adding 4% PFA (1 ml/35 mm dish) and incubate for 10 min at room temperature.
4. Discard the PFA into a toxic waste container and wash three times with 1 ml of PBS for 5 min each.
5. Permeabilize the cells by adding cold methanol (1 ml/35 mm dish) and incubate at −20°C for 6 min.
6. Discard the methanol into a toxic waste container and wash three times with 1 ml of PBS for 5 min each.
7. Remove the PBS and mark the border of the dish with a hydrophobic pen.

8. Block by adding PBS containing 5% goat serum (100 µl/35 mm dish) for 30 min at room temperature.
9. Remove the blocking solution and incubate the cells with the primary antibodies (100 µl/35 mm dish) (two-color stain), at the appropriate dilution (Pax7 (1:10) + MyoD (1:50)) and (MyoG (1:20) + MyHC (1:20)) diluted in sterile filtered PBS/BSA 4%) overnight in a humidified chamber at 4°C.
10. Remove the primary antibody solution and wash three times with 1 ml of PBS for 5 min each.
11. Incubate the cells with the secondary antibodies (100 µl/35 mm dish) (goat anti-rabbit Alexa Fluor 488 (1:1000) and goat anti-mouse Alexa Fluor 555 (1:1000) diluted in sterile filtered PBS/1% BSA) for 1 h at room temperature in a humified chamber in the dark.
12. Remove the secondary antibody solution and wash three times with 1 ml of PBS for 5 min each.
13. Counterstain the nuclei with Hoechst solution (100 µl/35 mm dish) for 5 min at room temperature in the dark.
14. Remove the Hoechst solution and wash three times with 1 ml of PBS for 5 min each.
15. Mount the coverslips with Vectashield mounting medium (30 µl/35 mm dish).
16. Analyze under a fluorescence microscope.

4 Notes

1. The 35 mm dishes will be needed 1 day following isolation.
2. It is important to use a pipette to remove the supernatant, as the pellet is not stable at this step.
3. Position the tubes horizontally, making sure they are completely submerged under water.
4. An adequately digested muscle will pass through the strainers easily. If necessary, the muscle digest can be passed again through another strainer to get rid of most of the cell debris and undigested muscle.
5. At this step, you can combine the pellets if more than one mouse is used. On average, the number of cells released after digestion from three 4- to 8-week-old mice is around 10^7 cells.
6. On average, the pre-plating steps remove around 50% of strongly adherent cells such as fibroblasts.

7. Both PBS with or without $CaCl_2$ can be used. However, using PBS without $CaCl_2$ will facilitate detachment.
8. If purity is of utmost importance (e.g., 99%–100% purity), the dishes should not be kept on ice for longer than 30 min. Nevertheless, the dishes can be kept on ice for up to 1 h to increase cell yield, with purity still high at 90%.
9. The original dishes (from day 1 of the isolation procedure) containing heterogeneous muscle cells can be kept in culture and used for further rounds of ICTs for up to day 3 of culture.
10. On average $1.5–2 \times 10^4$ SCs can be collected after 30 min on ice at 24 h of culture of the original dish. A further 4×10^4 SCs can be collected after the second or third ICT at 48 or 72 h of culture.
11. At this density, the SC proliferate up to 3 days after which they begin to differentiate. Increasing the plating density will accelerate the time it takes for the myoblasts to differentiate and fuse into myotubes.
12. The number of SCs obtained from human skeletal muscle biopsies may vary depending on the age of the donor, with a smaller number of cells typically obtained from older donors.
13. On average, from 1 g of human skeletal muscle biopsy, the number of muscle mononuclear cells obtained after digestion is around 2×10^6 cells.
14. To enrich the SCs present within the dish containing heterogeneous muscle cells on day 1 after isolation, it may be necessary to grow the heterogeneous culture for an additional 3–4 days before proceeding to ICT (when the culture reaches a confluence of about 80%).
15. On average 1×10^4 SCs can be collected after the first ICT. A further 2×10^4 SCs can be collected after the second or third ICT at 48 or 72 h of culture.
16. Human SCs proliferate slower than mouse SCs with a doubling time of 46 h.

Acknowledgments

This work was supported by a grant from The Dutch Duchenne Parent Project NL (DPP NL) to B.L.O.; research grants from Parent Project Italy (PP, Italy) to M.B.; and grants from the University of Rome (# RP11715C7D238352, RM118164275C7EBE, and RM11916B7E20311C to M.B. and # AR11715C7F9E158E, AR11816436905518, and AR11916B7E2A7B64 to A.B.). This protocol describes the method used in Benedetti et al. (2021).

Conflict of Interests The authors declare no competing financial or non-financial interests.

Ethics Approval and Consent for Publication All procedures involving mice were approved by the Italian Ministry for Health and were conducted according to the EU regulations and the Italian Law on Animal Research. Ethical Approval ID: 82945.19, valid until 18/01/2023. All patients gave informed consent to undergo intraoperative skeletal muscle biopsy and to publish the clinical and laboratory data obtained.

References

1. Mauro A (1961) Satellite cell of skeletal muscle fibers. J Biophys Biochem Cytol 9:493–495
2. Chang NC, Rudnicki MA (2014) Satellite cells: the architects of skeletal muscle. Curr Top Dev Biol 107:161–181
3. Wang YX, Dumont NA, Rudnicki MA (2014) Muscle stem cells at a glance. J Cell Sci 127(Pt 21):4543–4548.
4. Danoviz ME, Yablonka-Reuveni Z (2012) Skeletal muscle satellite cells: background and methods for isolation and analysis in a primary culture system. Methods Mol Biol 798:21–52
5. Keire P, Shearer A, Shefer G, Yablonka-Reuveni Z (2013) Isolation and culture of skeletal muscle myofibers as a means to analyze satellite cells. Methods Mol Biol 946:431–468
6. Syverud BC, Lee JD, VanDusen KW, Larkin LM (2014) Isolation and purification of satellite cells for skeletal muscle tissue engineering. J Regen Med 3(2):117
7. Gharaibeh B, Lu A, Tebbets J, Zheng B, Feduska J, Crisan M, Peault B, Cummins J, Huard J (2008) Isolation of a slowly adhering cell fraction containing stem cells from murine skeletal muscle by the preplate technique. Nat Protoc 3(9):1501–1509
8. Sherwood RI, Christensen JL, Conboy IM, Conboy MJ, Rando TA, Weissman IL, Wagers AJ (2004) Isolation of adult mouse myogenic progenitors: functional heterogeneity of cells within and engrafting skeletal muscle. Cell 119(4):543–554
9. Montarras D, Morgan J, Collins C, Relaix F, Zaffran S, Cumano A, Partridge T, Buckingham M (2005) Direct isolation of satellite cells for skeletal muscle regeneration. Science 309(5743):2064–2067
10. Pasut A, Oleynik P, Rudnicki MA (2012) Isolation of muscle stem cells by fluorescence activated cell sorting cytometry. Methods Mol Biol 798:53–64
11. Chapman MR, Balakrishnan KR, Li J, Conboy MJ, Huang H, Mohanty SK, Jabart E, Hack J, Conboy IM, Sohn LL (2013) Sorting single satellite cells from individual myofibers reveals heterogeneity in cell-surface markers and myogenic capacity. Integr Biol (Camb) 5(4):692–702
12. Fukada S, Higuchi S, Segawa M, Koda K, Yamamoto Y, Tsujikawa K, Kohama Y, Uezumi A, Imamura M, Miyagoe-Suzuki Y et al (2004) Purification and cell-surface marker characterization of quiescent satellite cells from murine skeletal muscle by a novel monoclonal antibody. Exp Cell Res 296(2):245–255
13. Liu L, Cheung TH, Charville GW, Rando TA (2015) Isolation of skeletal muscle stem cells by fluorescence-activated cell sorting. Nat Protoc 10(10):1612–1624
14. Blanco-Bose WE, Yao CC, Kramer RH, Blau HM (2001) Purification of mouse primary myoblasts based on alpha 7 integrin expression. Exp Cell Res 265(2):212–220
15. Benedetti A, Fiore PF, Madaro L, Lozanoska-Ochser B, Bouche M (2020) Targeting PKCtheta promotes satellite cell self-renewal. Int J Mol Sci 21(7):2419
16. Benedetti A, Cera G, De Meo D, Villani C, Bouche M, Lozanoska-Ochser B (2021) A novel approach for the isolation and long-term expansion of pure satellite cells based on ice-cold treatment. Skelet Muscle 11(1):7
17. Benedetti A, Cera G, De Meo D, Villani C, Bouche M, Lozanoska-Ochser B (2021) A simple method for the isolation and in vitro expansion of highly pure mouse and human satellite cells. Bio Protoc 11(23):e4238
18. Fiore PF, Benedetti A, Sandona M, Madaro L, De Bardi M, Saccone V, Puri PL, Gargioli C, Lozanoska-Ochser B, Bouche M (2020) Lack of PKCtheta promotes regenerative ability of muscle stem cells in chronic muscle injury. Int J Mol Sci 21(3):932

Check for updates

Chapter 3

3D Bioprinting Engineered Vascularized Skeletal Muscle Tissue

Anna Tsukerman, Eliana O. Fischer and Shulamit Levenberg

Abstract

3D bioprinting of vascularized tissues integrates additive manufacturing and tissue engineering to fabricate high-resolution, three-dimensional constructs. A key aspect of this method is embedding cells within a bioink, which provides structural support while encouraging cellular interaction during printing and subsequent cultivation. Commonly used bioinks are composed of extracellular matrix (ECM)-derived materials, mimicking the native tissue environment. Here, we describe a 3D bioprinting approach utilizing a collagen–fibrin bioink, termed ECM mix. This bioink formulation provides the mechanical strength required for skeletal muscle maturation while leveraging the proangiogenic properties of fibrin to support the formation of vascular networks. Printing is conducted at room temperature, and the constructs are incubated under physiological conditions to support tissue maturation.

Keywords 3D bioprinting, Vascular network, Skeletal muscle tissue, Tissue engineering, Regenerative medicine, Bioinks, Extracellular matrix

1 Introduction

3D bioprinting of vascularized tissue is anticipated to have a major impact on regenerative medicine, tissue transplantation, drug testing, and disease modeling. As a subdivision of tissue engineering, bioprinting shares its fundamental components, including biocompatible support materials (bioinks), living cells, and growth factors guiding differentiation pathways [1–3]. Among the efforts to fabricate different tissues, bioprinting skeletal muscle tissue is a rapidly advancing platform aiming to address the global organ shortage by producing large-scale, highly organized, and functional tissues [4, 5]. To achieve this goal, various bioprinting strategies, including stereolithography (SLA), digital light processing (DLP), and extrusion, are employed to form intricate 3D tissues [6, 7]. Extrusion-based bioprinting has emerged as the dominant technique for fabricating skeletal muscle

The authors Anna Tsukerman and Eliana O. Fischer contributed equally.

Joe Quadrilatero (ed.), *Myogenesis and Skeletal Muscle Biology*, Methods in Molecular Biology, vol. 3057,
https://doi.org/10.1007/978-1-0716-5396-8_3,

tissue, initially demonstrated by Atala et al. in 2018 [3, 8–9], due to its ability of forming continuous filamentous structures mimicking skeletal muscle fibers at a sufficient resolution and a relatively low expense [3, 10, 9]. A major challenge in extrusion-based bioprinting is maintaining intra-printing structural integrity until cross-linking occurs. Consequently, support baths were introduced, with Feinberg et al.'s development in 2015 of Freeform Reversible Embedding of Suspended Hydrogels (FRESH) using a gelatin-based matrix as a notable example. In this work, we employ "CarGrow," the K-carrageenan-based support bath that shares similar self-healing properties and tissue support during the printing and cultivation periods [11].

Here, we present an extrusion-based method of bioprinting vascularized skeletal muscle, with a spotlight on clinical applications where nutrient supply is crucial for tissue survival and integration. Our approach enables synchronous skeletal muscle maturation and microvasculature formation. The printed vascularized skeletal muscle tissue utilized human-derived cells suspended in ECM-based bioink composed of collagen and fibrin at a 2:1 ratio, further emphasizing its suitability for future clinical applications [9].

2 Materials

Sketch a 3D computer-aided design (CAD) for the bioprinted vascularized skeletal muscle tissue. Culture skeletal muscle cells, endothelial cells, and support cells prior to printing. Simultaneously, prepare the buffers, bioink, support bath, and culture media. Perform the bioprinting process while using an extrusion-based 3D bioprinter.

2.1 Construct Design

1. Create a 3D CAD design of the desired printed structure. For example, sketch a 6 mm × 6 mm × 1 mm cube using CAD software, such as SolidWorks (*see* **Notes 1 and 2**).
2. Use external slicer software to prepare the design for layer-by-layer printing. Set the infill density to 50% with an aligned pattern.
3. Define the layer height based on a 20% layer overlap, adjusted to the nozzle diameter (in this case, the design was based on a 0.22 mm, 27G needle) (*see* **Notes 3–6**).
4. Export the design as an STL file.
5. Open the STL file in the printer software and upload the slicing settings.

2.2 Material and Printing Parameter Selection

1. CarGrow support bath [10]: 1.5% kCar solution in 100 mM KCl, paraffin oil, Span 80, 70% ethanol. For two wells of a six-well plate, 10 mL of condensed support bath is required (*see* **Notes 7 and 8**).

2. ECM mix bioink preparation: Mix CollMA bioink, consisting of recombinant human collagen methacrylate (rhCollMA), crosslinked by lithium phenyl-2,4,6-trimethylbenzoylphosphinate (LAP), with polyethylene oxide (PEO) (Table 1), serving as a porogen. The abovementioned mixture is next mixed with fibrinogen at a 2:1 ratio to form an ECM mix [12].
3. rhCollMA mix solution:
   Immediately after preparing the PEO–LAP solution (Table 2), ensure minimal light exposure by wrapping it with aluminum foil.
   Create a 1× phosphate buffer solution by diluting the 10× phosphate buffer solution with deionized water at a 1:9 ratio (Table 3).
   Prior to printing, pipette the hydrogel vigorously while avoiding air bubbles. For rhCollMa–PEO–LAP crosslinking, use a

Table 1
Polyethylene oxide (PEO) porogen solution preparation

Reagents	Amount
PEO 1.6% (w/v)	160 mg
Phenol-red-free DMEM	10 L

Table 2
Photo-initiator solution (PEO–LAP) preparation

Reagents	Amount
LAP 0.2% (w/v)	6 mg
PEO solution (Sect. 2.2.2)	3 mL

Sterilize the PEO–LAP solution by passing it through a 0.22 μm filter and store it at 4°C until usage (*see* Note 9)

Table 3
Preparation of phosphate buffer solution (×10) for pH neutralization of the rhCollMA bioink stock [13]

Reagents	Amount
Na_2HPO_4	5.495 g
NaH_2PO_4	1.55 g
NaCl	30 mg
Deionized water	50 mL

Sterilize the buffer solution by passing it through a 0.22 μm filter and store it at room temperature

450 nm light source with an intensity of ~3 mW/cm^2, alternatively, use the bioprinter's UV curing tool (Table 4).

4. Fibrinogen solution (Table 5): Dilute the fibrinogen with the fibrinogen buffer to reach a concentration of 30 mg/mL.
5. ECM mix bioink (Table 6).

Table 4
Neutralization and dilution of the acidic rhCollMA to create a concentration of 5 mg/mL at a final volume of 1 mL

Reagents	Stock concentration	Amount
Step 1: rhCollMA at a concentration of 10 mg/mL		
rhCollMA	13.9 mg/mL	359.7 μL
Phosphate buffer	×10	50 μL
Phosphate buffer	×1	90.3 μL
Step 2: rhCollMA at a final concentration of 5 mg/mL		
rhCollMA (step 1)	10 mg/mL	500 μL
PEO–LAP	–	500 μL
After mixing the rhCollMa–PEO–LAP components (step 2), ensure minimal light exposure by wrapping it with aluminum foil (*see* **Note 10**)		

Table 5
Preparation of fibrinogen buffer

Reagents	Amount
Sodium chloride dihydrate	7 g
Trisodium citrate dihydrate	2.94 g
Glycine	9 g
Arginine hydrochloride	20 g
Calcium chloride dihydrate	0.15 g
Double-distilled water	1 L
Adjust the pH to 7.0 and store in aliquots at ≤ −30°C. Once thawed, store at 2–8°C for up to 1 month	

Table 6
Preparation of ECM mix bioink

Reagents	Amount
5 mg/mL rhCollMA-PEO-LAP (Sect. 2.2.3)	1 mL
30 mg/mL fibrinogen (Sect. 2.2.4)	0.5 mL

6. Cells:
   1. Prepare proliferation medium for the endothelial and support cells, and skeletal muscle cells.
   2. Prepare a skeletal muscle differentiation medium: Dulbecco's modified Eagle medium (DMEM), 5% horse serum, 1% Glutamax, and 1% penicillin–streptomycin solution.
   3. Culture separately the human umbilical vein endothelial cells (HUVEC), pericytes (support cells), and skeletal muscle cells in 75 cm^2 or 150 cm^2 cell culture flasks (*see* **Note 11**). Culture in an incubator under physiological conditions of 5% CO_2 at 37°C. Replace the medium every 2–3 days.

3 Methods

Throughout the printing process, perform all steps in the biological hood unless specified otherwise.

3.1 Bioprinter Preparation

1. Pre-sterilize the required equipment by autoclaving the syringes and needles in advance and incubate the plungers within 70% ethanol for at least 1 h (*see* **Note 12**).
2. Sterilize the printing compartment, including the bioprinting head and arm, syringe tools, and adaptors.
3. Upload the relevant printer container (six-well plate) and the STL files to the bioprinter.

3.2 Support Bath Preparation

1. Prepare CarGrow support bath via a water/oil emulsion method. Heat 1.5% kCar solution in 100 mM KCl with paraffin oil containing Span 80. Then homogenize, cool, and centrifuge to collect microgels. Discard the supernatant and replace the oil phase with ethanol, vortex, wash with KCl, and filter to remove aggregates before storage at 4°C. Prior to use, sterilize the support bath in ethanol, wash with sterile KCl, and incubate with up to 40 mL phenol-red-free medium with 5 U/mL of thrombin (*see* **Notes 13 and 14**).
2. Centrifuge at 3000 x *g* for 5 min and discard the supernatant. Flip the tube to ensure a condensed texture (Fig. 1A). With a positive-displacement pipette, transfer 5 mL support bath to the six-well plate and level the bath by tapping the plate several times (Fig. 1B).

3.3 Cell-Laden Bioink Preparation

1. Prepare cellular suspension containing 4×10^6/mL endothelial cells with 16×10^6/mL pericytes in a centrifuge tube.

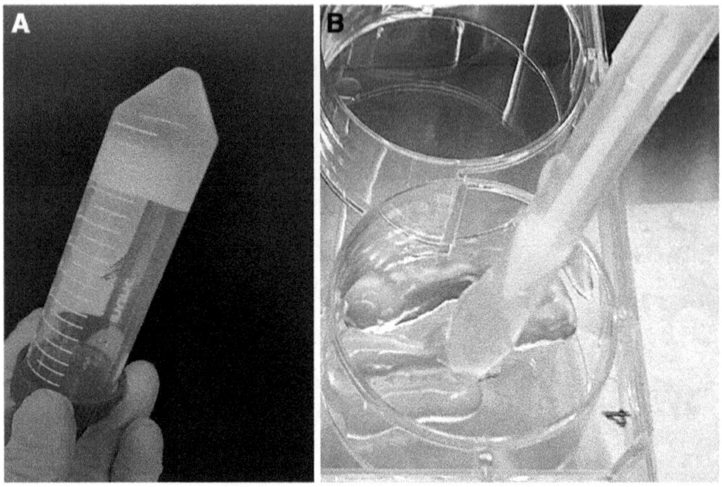

Fig. 1 Granular support bath. (**A**) The condensed form of the support bath. (**B**) Positive displacement of the support bath into a six-well plate

Table 7
Cell type and the required cell concentration

Cell type	Required number of cells per mL
Endothelial cells (HUVECs)	$4–5 \times 10^6$
Support cells (pericytes)	$16–20 \times 10^6$
Skeletal muscle cells	$12–15 \times 10^6$

In another centrifuge tube, prepare a suspension of 15×10^6/mL skeletal muscle cells. See Table 7.

2. Centrifuge at $200 \times g$ for 4 min to obtain a cell pellet. Aspirate the supernatant medium of each centrifuge tube.
3. Resuspend each cell pellet in 0.75 mL ECM-mix bioink containing rhCollMA–PEO–LAP and fibrin (*see* **Note 15**).
4. Prepare a 3 mL printing cartridge with a plunger inserted up to the orifice of the cartridge.
5. Transfer the cell-laden bioink mixture into the cartridge by filling it from the nozzle opening, using a positive-displacement pipette while avoiding bubbles (*see* **Note 16**). Attach a 0.22 mm internal diameter nozzle to the cartridge. To maintain sterility upon transfer from the hood to the bioprinter, place the loaded cartridge with its nozzle in a 50 mL centrifuge tube.
6. Assemble the loaded cartridge into the 3 mL ambient pneumatic dispensing tool of the 3D extrusion-based bioprinter.

7. Calibrate the bioink settings (PSI, speed) for each syringe.
8. Modify the STL file based on pressure and speed parameters and send it to the printer.
9. For rhCollMA–PEO–LAP crosslinking, immediately after printing, expose the printed tissue to a 405 nm light source with an intensity of ~3 mW/cm^2 for 30 s. Alternatively, cure while printing and define the material settings accordingly.
10. Incubate the constructs at 37°C and 5% CO_2 for 30 min to enhance thrombin crosslinking of the fibrinogen component to fibrin.
11. Add proliferation medium for both endothelial cells and skeletal muscle cells at a 1:1 ratio, while leaving the constructs in the support bath wells.
12. A significant portion of the support bath will gradually liquefy over the 24 h following printing. Gently aspirate the liquefied support bath mixed with medium distant from the printed cells. Next, transfer the printed construct to a new culture plate using a spatula, leaving the remaining support bath in the original well. Add proliferation medium as described above for an additional 24 h (*see* **Notes 17 and 18**).
13. Replace the medium with a mixture of the proliferation medium of endothelial cells and differentiation skeletal muscle medium in a 1:1 ratio. Every 2–3 days, replace the medium until the experiment is terminated (*see* **Note 19**).

4 Notes

1. Before proceeding to cellular printing, conduct a material-only printing session preferably with fluorescent beads that will enable structural imaging.
2. When designing the printed structure, the diffusion limit of nutrients must be considered.
3. Ensure nozzle height calibration to prevent collision with the bottom of the printing well. Verify the height by printing a line with an empty cartridge before proceeding to the cellular print.
4. Using fluorescently labeled cells will enable cellular localization post-printing and monitoring of cellular morphogenesis.
5. At the time of geometry design, a layer overlap of at least 20% will create stable structures that can safely be extracted from the support bath.

6. Constructs composed of alternate layers of skeletal muscle and vasculature were designed in this protocol; however, all three cellular lineages can be mixed and printed using a single cartridge.
7. Other support baths such as the gelatin-based FRESH support bath can be used. Take into account the immediate liquefication at 37°C and the subsequent effect on the printed structures.
8. Up to four constructs can be printed in a single well of a six-well plate.
9. Store the PEO–LAP solution according to the LAP manufacturer's instructions.
10. If the bioink includes non-sterile components, filter with a 0.22 μm filter before use. Alternatively, powders can be sterilized via UV exposure prior to mixing with the sterile components
11. Other endothelial and support cells can be used as an alternative to HUVECs and pericytes. In that case, an additional step of calibrating the support cell–endothelial cell ratio would be required.
12. Printing equipment should be used as disposables if possible.
13. The speed and duration of the support bath centrifugation have a significant effect on the condensation of the bath. If the bath displaces easily when tilted in the falcon, centrifuge again until a condensed form is visible.
14. When using a different bioink, the relevant crosslinker should be added to the support bath. Calculate the concentration per mL and multiply it by volume contained in the centrifuge tube (support bath + medium).
15. A bioink volume of 1.5 ml is sufficient for printing approximately 16 constructs.
16. Make sure that the plunger is moving downward at the time of printing to ensure bioink extrusion. Alternatively, image the printed structures and locate fluorescently labeled cells or fluorescent beads.
17. Construct extraction can be performed on the day of printing, after cross-linking occurs.
18. An alternative approach to construct extraction relies on gently pipetting PBS into the bath, distant from the printed cells, and aspirating the liquefied support bath. This step should be repeated until the printed structures are clearly distinguishable.

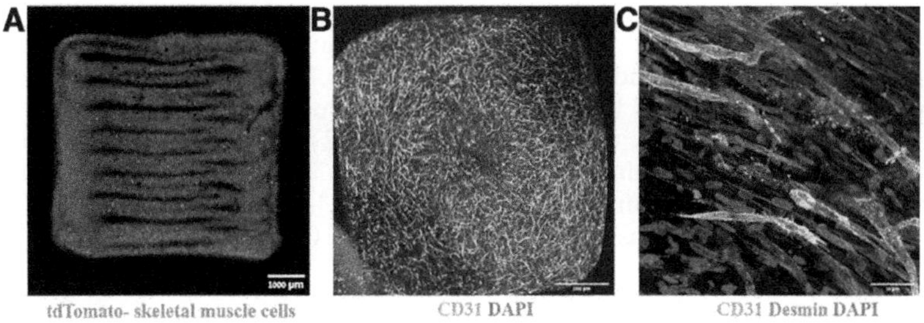

Fig. 2 Confocal microscopy representative images of final bioprinted vascularized skeletal muscle construct. (**A**) Fluorescently labeled tdTomato skeletal muscle cells. Scale bar = 1000 μm. (**B**) Vascular networks of endothelial cells (HUVECs) and unstained human pericytes (HPs). Constructs were stained for CD31 (green) and nuclei (DAPI; blue). Scale bar = 1000 μm. (**C**) Immunofluorescent staining of skeletal muscle cells (desmin, red), HUVECs (CD31, green), unstained pericytes, and nucleus (DAPI, blue). Scale bar = 50 μm

19. At the end of the cultivation period, different analyses can be performed, such as immunofluorescent staining (Fig. 2), RNA extraction, live–dead assays, etc.

Funding

This research received funding from the European Research Council (ERC) under the European Union's Horizon 2020 Research and Innovation Programme (VesselNet' grant agreement no. 818808).

References

1. Bliley JM, Shiwarski DJ, Feinberg AW (2022) 3D-bioprinted human tissue and the path toward clinical translation. Sci Transl Med 14, eabo7047.
2. Bishop ES et al (2017) 3-D bioprinting technologies in tissue engineering and regenerative medicine: current and future trends. Genes Dis 4:185–195
3. Samandari M et al (2022) Bioinks and bioprinting strategies for skeletal muscle tissue engineering. Adv Mater 34:1–29
4. Mandrycky C, Wang Z, Kim K, Kim DH (2016) 3D bioprinting for engineering complex tissues. Biotechnol Adv 34:422–434
5. Atala A (2007) Engineering tissues, organs and cells. J Tissue Eng Regen Med 1:83–96
6. Zennifer A, Manivannan S, Sethuraman S, Kumbar SG, Sundaramurthi D (2021) 3D bioprinting and photocrosslinking: emerging strategies & future perspectives. Biomater Adv 134:112576
7. Szklanny AA, et al (2021) 3D bioprinting of engineered tissue flaps with hierarchical vessel networks (VesselNet) for direct host-to-implant perfusion. Adv Mater 33, e2102661. https://doi.org/10.1002/adma.202102661
8. Hinton TJ, et al (2015) Three-dimensional printing of complex biological structures by freeform reversible embedding of suspended hydrogels. Sci Adv 9(1–10), e1500758.
9. Kim JH et al (2020) Neural cell integration into 3D bioprinted skeletal muscle constructs accelerates restoration of muscle function. Nat Commun 11:1–12
10. Machour M et al (2022) Print-and-grow within a novel support material for 3D bioprinting and post-printing tissue growth. Adv Sci 9:e2200882

11. Choi Y (2019) A 3D cell printed muscle construct with tissue-derived bioink for the treatment of volumetric muscle loss. Biomaterials 206:160–169.
12. Fischer EO, Tsukerman A, Ikushima K, et al. (2026) Bioprinted vascularized soft-tissue flaps with an integrated arterial-venous loop. Cell Biomater 0:0. https://doi.org/10.1016/j.celbio.2026.100385
13. Machour M, Szklanny AA, Levenberg S (2022) Fabrication of engineered vascular flaps using 3D printing technologies. J Vis Exp 183, e63920. https://doi.org/10.3791/63920

Chapter 4

Engineering Functional Skeletal Muscle Tissues from Human Induced Pluripotent Stem Cells

Amber Detwiler, Alastair Khodabukus and Nenad Bursać

Abstract

In vitro cell culture models of human skeletal muscle offer a valuable platform for studies of pathology and discovery of new therapeutics. However, traditional two-dimensional cell culture models suffer from limited durability and challenges with assessment of skeletal muscle contractile function. Furthermore, finite proliferation potential and scarce availability of patient derived primary myoblasts limit experimental throughput when using these cells. Here, we describe a method for using human induced pluripotent stem cells (hiPSCs) to generate functional three-dimensional skeletal muscle tissues. Briefly, transient overexpression of Pax7 in hiPSC-derived paraxial mesoderm cells permits efficient formation of myogenic progenitor cells (iMPCs), which are then combined with hydrogel matrix to form cylindrically shaped skeletal muscle (iSKM) "myobundles." After 2 weeks of differentiation, iSKM myobundles contain aligned and densely packed multinucleated myotubes that, in response to electrical stimulation, exhibit robust calcium transients and positive force–frequency and force–length relationships. The methods detailed here have been successfully utilized to model and study limb-girdle muscular dystrophy type R2/2B and can be applied for studies of skeletal muscle regeneration and testing of candidate gene and cell therapies for disease. Collectively, iSKM myobundle engineering is a powerful in vitro methodology expected to significantly advance human skeletal muscle research and therapy.

Keywords Skeletal muscle, Human induced pluripotent stem cells, Tissue engineering, Myopathy, Disease modeling, Hydrogel

1 Introduction

Impaired skeletal muscle function is a defining feature of numerous genetic and acquired myopathies including muscular dystrophies [1], inflammatory myopathies [2], and neuromuscular diseases [3]. Unfortunately, when compared to human pathology, animal models of skeletal muscle disease are significantly less severe, which greatly limits their utility in therapeutic development [4–6]. Consequently, the generation of in vitro functional human skeletal muscle models can provide unique opportunities to study fundamental disease biology and effectively screen new candidate therapeutics.

Joe Quadrilatero (ed.), *Myogenesis and Skeletal Muscle Biology*, Methods in Molecular Biology, vol. 3057,
https://doi.org/10.1007/978-1-0716-5396-8_4,

Existing in vitro models of human skeletal muscle [7–10] commonly utilize primary myoblasts that enable patient-specific studies. Although this approach has certain advantages, it relies upon tissue biopsies that are notoriously difficult to obtain from patients with rare myopathies. Furthermore, primary myoblasts exhibit limited proliferative capacity [11] and display a progressive loss of myogenic potential with expansion in vitro [12]. While immortalization can circumvent these limitations, the immortalization process may alter cellular biology [13], induce genetic alterations [13], and overlook the potential significance of senescence in dystrophic severity [14]. In contrast, human induced pluripotent stem cells (hiPSCs) [15, 16] represent a promising cellular alternative for in vitro disease modeling due to their potential for unlimited expansion and maintenance of pathological cell phenotype [17–19]. Myogenic progenitors have been successfully derived from hiPSCs using small molecule [20] or genetic reprogramming [21–23] techniques, with the latter being more rapid and able to generate a higher yield and purity of myogenic progenitors. After transplantation in vivo, hiPSC-derived myogenic cells have been shown to fuse with host myofibers and improve skeletal muscle function in dystrophic mouse models [24–27], underscoring their potential as a therapeutic surrogate for primary myoblasts.

The majority of in vitro skeletal muscle models employ two-dimensional (2D) cell culture [7–10], which enables rapid assessment of myotube fusion and sarcomere organization. However, 2D models undergo myotube detachment within several days of differentiation and are not amenable to measurements of contractile function (i.e., force generation and fatigue) [28]. To overcome these shortcomings, we previously developed a 3D tissue-engineered human skeletal muscle ("myobundle") platform using either primary [29–36] or hiPSC-derived [23, 37] MPCs that generate calcium transients and tetanic contractions in response to electrical and biochemical stimuli. Importantly, myobundles display physiologically relevant responses to drug treatments [29, 31], exercise [33], and both genetic and acquired diseases [30, 33, 37], supporting their utility as an in vitro 3D model of human skeletal muscle.

Here, we describe an optimized hiPSC-to-iMPC differentiation protocol involving transient Pax7 overexpression in hiPSC-derived paraxial mesoderm cells [23]. The robustness of this protocol has been validated across multiple hiPSC lines in both 2D and 3D culture [23, 37], consistently producing cross-striated and multinucleated myotubes and a Pax7+ satellite-cell pool. Within our 3D culture, iMPCs mature into functional skeletal muscle (iSKM) myobundles, which generate contractile twitch and tetanic forces, calcium transients, and progressive hypertrophy over a 2- to 4-week differentiation period [23]. Recently, these methods enabled the creation of the first hiPSC-derived model of limb-girdle muscular dystrophy type R2/2B [37], which recapitulated key disease features and pharmaceutical responses found in vivo. Furthermore, both primary and hiPSC-derived myobundles successfully engraft and survive upon transplantation in immunocompromised rodents [23, 36], laying a foundation for development of future tissue replacement therapies. These methods can be extended to other patient-specific MPCs for studies of disease mechanisms and testing of therapeutics.

2 Materials

2.1 Manufacturing of Tissue Molds and Frames

Methods for fabrication of myobundle molds and frames were reproduced and modified from our previously published cardiac tissue manufacturing protocol [38].

1. Polytetrafluoroethylene (PTFE).
2. Computer numerical control (CNC) router.
3. SYLGARD™ 184 Silicone Elastomer Kit (polydimethylsiloxane, PDMS).
4. Vacuum desiccator.
5. Oven set to 60°C.
6. Velcro®, soft "loop" side (*see* **Note 1**).
7. Laser cutter.
8. Ultrasonic bath.
9. 70% ethanol.
10. Petri dishes.
11. Razor blades.

2.2 Lentivirus Culture and Collection

All components needed for HEK-293FT culture should be prepared in a sterile fashion and filtered using a 0.22-μm filter before use.

1. HEK-293FT cells.
2. HEK-293FT growth medium: High-Glucose Dulbecco's Modified Eagle Medium (HG DMEM), 10% (v/v) fetal bovine serum (FBS), 1% (v/v) penicillin–streptomycin.
3. Gelatin-coated 10-cm cell culture plates.
4. pRRL-TetO-Pax7-IRES-EGFP DNA expression and FUW-M2rtTA reverse tetracycline transactivator lentiviral plasmids (*see* **Note 2**).
5. psPAX2 lentiviral packaging plasmid (*see* **Note 2**).
6. pMD2.G lentiviral VSV-G-expressing envelope plasmid (*see* **Note 2**).
7. Lipofectamine™ 2000 Transfection Reagent.

8. Opti-MEM™ Reduced Serum Medium.
9. 0.45 μm cellulose acetate filter.
10. Lenti-X™ Concentrator.
11. −80°C freezer.
12. Incubator set to 37°C, 5% CO_2.
13. Temperature-controlled centrifuge set to 4°C.

2.3 iMPC Differentiation Cell Culture Media and Supplements

All components should be prepared in a sterile fashion and filtered using a 0.22-μm filter before use.

1. Doxycycline-Inducible Pax7/GFP Lentivirus and Reverse Tetracycline Transactivator Lentivirus collected in Sect. 3.2.
2. Human induced pluripotent stem cells (hiPSCs).
3. Six-well culture plates, treated.
4. mTESR™ Plus Medium (*see* **Note 3**).
5. Dulbecco's phosphate-buffered saline (DPBS) without $CaCl_2$ and $MgCl_2$.
6. 0.5 μM ethylenediaminetetraacetic acid disodium salt solution (EDTA).
7. ACCUTASE™ Cell Detachment Solution.
8. Matrigel® hESC-Qualified Matrix.
9. Essential 8™ Medium.
10. Essential 6™ Medium.
11. CryoStor® CS10 Freezing Medium.
12. ROCK inhibitor Y-27632.
13. CHIR99021.
14. Doxycycline (DOX).
15. Basic fibroblast growth factor (bFGF).
16. Trypsin-EDTA (0.25%).
17. Trypsin-EDTA (0.05%).
18. Neutralizing media: Dulbecco's Modified Eagle Medium/Nutrient Mixture F-12 (DMEM/F-12), 20% (v/v) FBS, with 40 μg/mL DNase I, 5 μM Y-27632 added immediately before use.
19. 40-μm cell strainer.
20. 30-μm cell strainer.
21. Sorting solution: Essential 6™ Medium, 1% (v/v) penicillin–streptomycin, with 10 μg/mL DOX, 10 μM Y-27632, 40 μg/mL DNase I added immediately before use.

22. Collection solution: FBS, 1% (v/v) penicillin–streptomycin with 10 μg/mL DOX, 10 μM Y-27632 added immediately before use.
23. Single-cell suspension of undifferentiated hiPSCs in sorting solution.
24. Flow cytometry cell sorter (*see* **Note 4**).
25. Matrigel® Growth Factor Reduced (GFR) Matrix.
26. T-75 tissue culture flasks, treated.
27. T-175 tissue culture flasks, treated.
28. Expansion media: Low-Glucose Dulbecco's Modified Eagle Medium (LG DMEM), 10% (v/v) FBS, 1% (v/v) penicillin–streptomycin, 400 ng/mL dexamethasone (DEX), 10 ng/mL epidermal growth factor (EGF), with 1 μg/mL DOX, 10 ng/mL bFGF added immediately before use.
29. Hemocytometer.
30. Incubator set to 37°C, 5% CO_2.
31. Temperature-controlled centrifuge set to 4°C.

2.4 3D iSKM Tissue Generation

All components should be prepared in a sterile fashion and filtered using a 0.22-μm filter before use.

1. Pluronic F-127: 0.2% (w/v) in deionized H_2O.
2. Dulbecco's phosphate-buffered saline (DPBS) without $CaCl_2$ and $MgCl_2$.
3. Dulbecco's Modified Eagle Medium/Nutrient Mixture F-12 (DMEM/F 12).
4. 20 million iMPCs/mL in expansion media.
5. Bovine fibrinogen.
6. Syringe, luer lock tip.
7. 0.22-μm filter syringe attachment.
8. Bovine thrombin: 50 U/mL in 0.1% (w/v) bovine serum albumin solution.
9. Matrigel® Growth Factor Reduced (GFR) Matrix.
10. Forceps.
11. 12-well culture plates, non-treated.
12. Sterile Petri dishes.
13. 3D growth media: LG DMEM, 10% (v/v) FBS, 1% (v/v) penicillin–streptomycin, 400 ng/mL DEX, 10 ng/mL EGF, with 1 μg/mL DOX, 1.5 mg/mL 6-aminocaproic acid (ACA) added immediately before use

14. 3D differentiation media: LG DMEM, 1% (v/v) N-2 Supplement (100×), Penicillin G (100 unit/mL), 2 mg/mL ACA, 50 mg/mL ascorbic acid 2-phosphate, 200 μM L-carnitine, 10 ng/mL biotin.
15. Triiodothyronine (T3).
16. SEL media: 3D differentiation media, with 0.1 μM smoothened agonist (SAG), 2 ng/mL EGF, 5 nM LY209314 added immediately before use.
17. Incubator set to 37°C, 5% CO_2.
18. Water bath set to 37°C.
19. Temperature-controlled centrifuge set to 4°C.
20. High-angle rocker set to ±30° tilt and 0.4 Hz rocking frequency.

3 Methods

3.1 Manufacturing of PDMS Tissue Molds and Velcro® Frames

3D engineered skeletal muscle tissues are formed within molds containing two semi-cylindrical wells (5 mm long, 1 mm diameter) (Fig. 1). Tissues are supported by a single-use Velcro® frame (7×7 mm^2, 1 mm wide rim) (Fig. 1) for easy removal from molds for further culture.

1. For generating tissue molds, design a master negative using 3D design software.
2. Machine polytetrafluoroethylene (PTFE) into negative master shape using a CNC router. Master negatives can also be 3D printed depending on design intricacy.

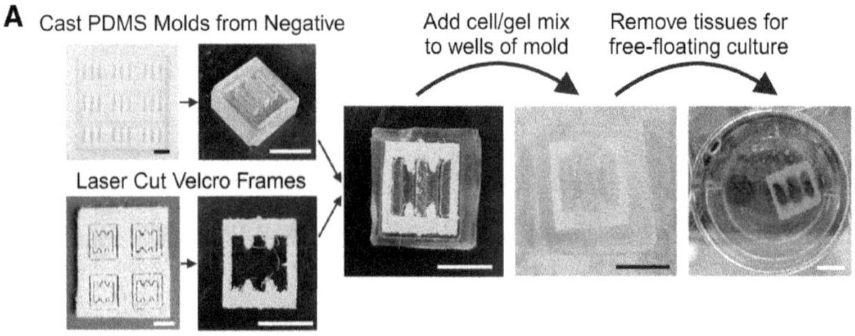

Fig. 1 Mold and frame system for 3D tissue-engineered iSKM myobundle generation. PDMS molds are cast from Teflon negatives and frames are laser-cut from the loop-side sheet of Velcro®. Assembled molds and frames are inoculated with hydrogel and iMPC mix. Tissues are gelled at 37°C before removal for culture in a standard 12-well plate. Scale bar = 5 mm

3. Prepare the polydimethylsiloxane (PDMS) by thoroughly mixing SYLGARD™ 184 Silicone Elastomer base and SYLGARD™ 184 Silicone Elastomer curing solution in a 10:1 ratio (w:w).
4. Transfer the PDMS to a vacuum desiccator and degas for 30 min until all bubbles are removed.
5. Place the PTFE master(s) face up in a Petri dish and pour the PDMS over the master until it is completely submerged. There should be ~2 mm of PDMS above the highest point of the master.
6. Transfer the dish containing the PDMS and master to a vacuum desiccator and degas for 30 min, ensuring that all bubbles are evacuated. Remaining bubbles may be removed with a needle or pipette tip.
7. Cure the PDMS in a 60°C oven for at least 4 h, ensuring that the dish is leveled. PDMS can also be cured at room temperature for several days if the master negative is made from a non heat-safe material.
8. After the curing is complete, remove the PDMS from the dish and carefully extract the PTFE master. Remove the excess PDMS that surrounds the mold using a sharp razor blade to produce a final PDMS mold.
9. Submerge the PDMS molds in 70% ethanol and sonicate in an ultrasonic bath for 90 min at 60°C. Store sonicated molds in 70% ethanol until needed (*see* **Note 5**).
10. For generating single-use Velcro® frames, design a frame template using 2D drawing software that is compatible with the laser cutter. Connect frames to the main sheet of Velcro® via thin bridges of uncut material to allow for easy removal after laser cutting.
11. Laser-cut frames out of Velcro®, ensuring that the laser settings do not burn the material. We have found that laser cutting Velco® with the soft/fuzzy side down works best.
12. Carefully separate individual frames by cutting the thin bridges that are holding the frames to the larger sheet of nylon material.
13. Approximately 24–48 h before myobundle making, the molds and frames can undergo final sterilization steps (steps 14–17 below).
14. Using sterile forceps, place the sonicated PDMS molds into a sterile Petri dish. Leave at least 3 mm of space between molds to facilitate eventual mold removal.
15. Soak Velcro® frames in 70% ethanol solution for at least 30 min before pouring into a sterile Petri dish.
16. Aspirate off excess ethanol from the molds and frames.

17. Leave the lids of the petri dishes slightly ajar to allow the molds and frames to dry overnight. The sash of the biosafety cabinet can be closed with UV turned on as an added layer of sterilization.

3.2 Preparation of Doxycycline-Inducible Pax7/GFP and Reverse Tetracycline Transactivator Lentivirus

All steps should be performed in a BSL-2 biosafety cabinet. All incubation steps should be performed in a standard cell culture incubator set to 37°C, 5% CO_2. Store lentivirus at −80°C in single-use aliquots to avoid freeze-thaw cycles.

1. 24 h before transfection, plate HEK-293FT cells on gelatin-coated 10-cm cell culture plates at approximately 7.5×10^4 cells/cm^2.
2. On day of transfection, change HEK media at least 3 h before transfection. HEKs should be 60%–80% confluent at time of transfection.
3. For transfection, combine diluted plasmids ("Mastermix A") and diluted Lipofectamine™ ("Mastermix B") in a 1:1 volume ratio before adding the entire 200 μL mix to one 10-cm cell culture plate. Adjust according to total number of 10-cm plates for transfection.
4. Generate Mastermix A by combining 10 μg pRRL-TetO-Pax7-IRES-EGFP plasmid with 5 μg psPAX2 plasmid and 2 μg pMD2.G plasmid. Add Opti-MEM™ to reach a final volume of 100 μL. Vortex briefly and incubate at room temperature for 5 min.
5. Generate Mastermix B by diluting 2 μL Lipofectamine 2000 in 98 μL Opti-MEM™. Vortex briefly and incubate at room temperature for 5 min.
6. Combine 100 μL Mastermix A and 100 μL Mastermix B. Vortex briefly and incubate at room temperature for 10–20 min.
7. Add mixture dropwise to the HEK plate, then rock plate gently to disperse.
8. Return cell culture plate to incubator (37°C, 5% CO_2).
9. Change HEK media 12–16 h later.
10. Check cells one day post-media change. If most cells are dead, harvest. If not, wait one more day.
11. When ready for harvest, collect HEK media and centrifuge at 300 × *g* for 5 min in a temperature-controlled centrifuge set to 4°C.
12. Filter supernatant in a 0.45-μm filter to remove cell debris.
13. Combine filtered supernatant with Lenti-X™ Concentrator in a 3:1 volume ratio and concentrate to 1/100 of the original volume per manufacturer's instructions.
14. Repeat Section 3.2 to generate the reverse tetracycline transactivator lentivirus, adjusting step 4 as follows, "Generate

Mastermix A by combining 10 μg FUW-M2rtTA plasmid with 5 μg psPAX2 plasmid and 2 μg pMD2.G plasmid."

3.3 Differentiation of hiPSCs into Myogenic Progenitor Cells (iMPCs)

All steps should be performed in a BSL-2 biosafety cabinet. All incubation steps should be performed in a standard cell culture incubator set to 37°C, 5% CO_2.

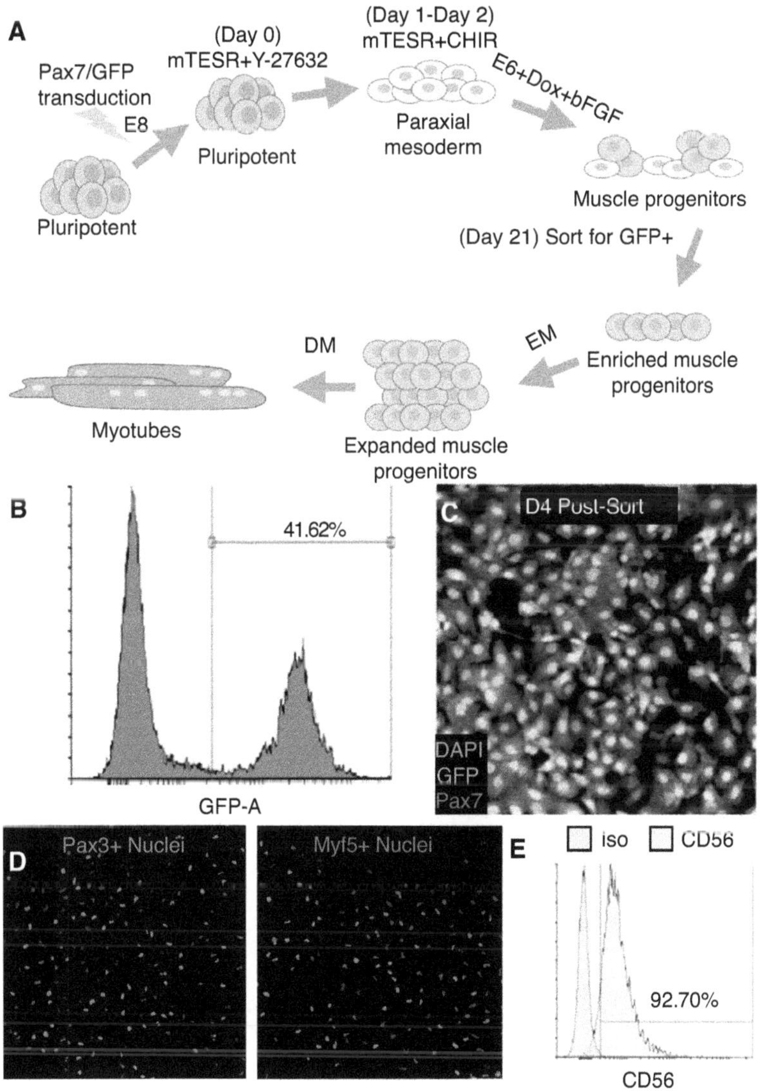

Fig. 2 Differentiation of myogenic progenitor cells (MPCs) from hiPSCs. (**A**) Schematic of hiPSC to iMPC to myotube differentiation procedure. Dox, doxycycline; EM, expansion media; DM, differentiation media. (**B**) Representative results of fluorescence-activated cell sorting (FACS) for GFP+ cells at day 21. (**C**) Representative immunostaining of Pax7 and GFP in expanded iMPCs 4 days post-sort. Scale bar = 50 μm. (**D**) Representative immunostaining of Pax3 (left) and Myf5 (right) during iMPC expansion. Scale bar = 100 μm. (**E**) Representative FACS results for CD56+ iMPCs. (Reproduced with permission from [23])

1. Maintain hiPSCs in mTESR™ Plus media, performing EDTA passages (*see* **Note 6**) as needed until ready to begin differentiation.
2. Once ready to begin iMPC differentiation (Fig. 2A), dissociate 70%–80% confluent hiPSC colonies into single cells with ACCUTASE™.
3. Suspend single cells in Essential 8™ medium and seed onto hESC-qualified Matrigel®-coated six-well plates (*see* **Note 7**) at a cell density of $8 \times 10^4/\text{cm}^2$.
4. 24 h after plating, transduce the hiPSCs with the dox-inducible Pax7 overexpression lentivirus and reverse tetracycline transactivator lentivirus generated in Sect. 3.2 (*see* **Note 8**).
5. Expand transduced hiPSCs in Essential 8™ media for at least 3 and up to 20 passages (*see* **Note 9**).
6. Day 0: At 70%–80% confluency of transduced hiPSCs, dissociate colonies into single cells with ACCUTASE™. Suspend single cells in mTESR™ supplemented with Y-27632 (5 μM) and seed onto hESC-qualified Matrigel®-coated six-well plates (*see* **Note 7**) at a density of 2.6×10^4 cells/cm^2.
7. Days 1–2: Feed daily with mTESR™ media supplemented with CHIR99021 (10 μM). Replace with fresh medium at the same time each day.
8. Days 3–5: Feed daily with E6 media supplemented with 1 μg/mL DOX. Replace with fresh medium at the same time each day.
9. Days 6–21: Feed daily with E6 media supplemented with 1 μg/mL DOX and 10 ng/mL bFGF.
10. Once cells reach day 21, they are ready for sorting by FACS for GFP (*see* **Note 10**).

3.4 Sorting of iMPCs

All steps should be performed in a BSL-2 biosafety cabinet. All incubation steps should be performed in a standard cell culture incubator set to 37°C, 5% CO_2.

1. Rinse day 21 iMPC differentiation plate twice with 1× DPBS.
2. Immediately apply 0.25% trypsin-EDTA and incubate the plate for 3–5 min at 37°C (7 min maximum).
3. Use tapping and a 10 mL serological pipette to break up clumps and loosely adherent cells.
4. Transfer cell suspension to equal volume of neutralizing media. Steps 2-4 may be repeated up to 2 more times until cell clumps are fully broken-up. Collagenase may be added to trypsin-EDTA at 2 mg/mL in cases of excess ECM deposition.

5. Pass suspension through a 40-μm cell strainer (*see* **Note 11**).
6. Centrifuge at 300 × *g* for 5 min in a temperature-controlled centrifuge set to 4°C.
7. Resuspend the cell pellet in 5 mL of sorting solution, pipetting seven times to mix.
8. Pass suspension through a 30-μm cell strainer and flush strainer with 1 mL sorting solution.
9. Store differentiated iMPC cell suspension, undifferentiated hiPSC cell suspension, and adequate volume of collection solution on ice until ready to sort.
10. Sort for GFP+ cells, using undifferentiated single-cell hiPSCs as a negative control. Expect a yield of approximately 40% GFP+ iMPCs (Fig. 2B).
11. Centrifuge sorted iMPCs at 300 × *g* for 5 min in a temperature-controlled centrifuge set to 4°C.
12. Resuspend in expansion media supplemented with 5 μM Y-27632.
13. Plate 4 × 10^4 iMPCs/cm^2 on GFR Matrigel®-coated flasks (*see* **Note 7**) in expansion media supplemented with 5 μM Y-27632.

3.5 Maintenance of iMPCs

All steps should be performed in a BSL-2 biosafety cabinet. All incubation steps should be performed in a standard cell culture incubator set to 37°C, 5% CO_2.

1. Change media the day after sorting using expansion media.
2. Once the flasks reach 80%–90% confluency, they are ready to passage. This is usually 4-6 days after sorting (*see* **Note 12**).
3. Aspirate media from the flasks and rinse with 1× DPBS.
4. Apply 0.05% trypsin-EDTA and place in the incubator for 3–5 min.
5. Once cells are free (use gentle tapping if cells are still stuck to flask), add equal volume of expansion media and collect cell suspension.
6. Count cells using a hemocytometer.
7. Centrifuge at 300 × *g* for 5 min in a temperature-controlled centrifuge set to 4°C.
8. Suspend in expansion media and replate at a 1:4 to 1:8 ratio on GFR Matrigel®-coated flasks (*see* **Note 7**).
9. After the iMPCs reach 80%–90% confluency, they are ready to be passaged for myobundle making (Fig. 2C–E).

3.6 Fabrication of 3D Tissue-Engineered "Myobundles"

All steps should be performed in a BSL-2 biosafety cabinet. All incubation steps should be performed in a standard cell culture incubator set to 37°C, 5% CO_2.

1. No more than 24 h before myobundle making, fully sterilized molds should be coated with Pluronic F-127 solution to decrease cell attachment.
2. Allow Pluronic F-127-filled molds to sit undisturbed at room temperature for 2–3 h before aspirating off.
3. Once molds are fully dry, gently press Velcro® frames into the molds with forceps, ensuring the soft, fuzzy side of the frame is face-up (Fig. 1).
4. Optionally, assembled frames and molds can be UV sterilized in the biosafety cabinet for 30 min before use.
5. Myobundle formation includes the combination of equal parts "gel solution" and "cell solution" before transferring to assembled mold/frame units. Unless otherwise specified, all remaining myobundle fabrication steps should be done on ice.
6. First determine the mass of fibrinogen needed to generate the desired volume of 20 mg/mL fibrinogen in 1× DPBS. Plan to prepare at least 300 μL extra to account for loss during sterile filtration. Aliquot the determined mass of fibrinogen.
7. Add the appropriate volume of 1× DPBS and gently pipette to mix.
8. Heat the solution at 37°C until completely dissolved (15–45 min) and filter with a luer lock–tipped syringe with a 0.22-μm filter attached (*see* **Note 13**).
9. To prepare the gel solution, combine GFR Matrigel®, fibrinogen solution, and DMEM-F12 media in a 2:2:1 volume ratio.
10. Divide gel solution into microtubes according to mold volume. Plan to form no more than four myobundles (two full molds) at a time due to the rapid cross-linking of fibrinogen by thrombin (*see* **Note 14**).
11. Repeat steps 3–7 of Sect. 3.5 to collect iMPCs for myobundle making.
12. Resuspend iMPCs at 20 million/mL in expansion media and keep on ice.
13. To prepare the cell solution, combine 92% iMPC cell suspension with 8% thrombin by volume.
14. Working quickly and with all reagents on ice, add equal volume of cell solution to a microtube of gel solution. Pipette several times to mix (avoiding bubbles) and immediately

transfer one myobundle's worth of the gel/cell mixture into one well of an assembled mold. Gently "paint" a small amount of the liquid mixture into the notches of the frame to ensure secure attachment after gelation.

15. Repeat step 14 without changing pipette tips to quickly distribute the remaining mixture into the second well of the mold, repeating further if more than two myobundles are being made at a time (*see* **Note 15**).
16. Repeat steps 14–15 to make remaining myobundles (Fig. 1).
17. Place the Petri dish containing the hydrogel-filled molds into a 37°C, 5% CO_2 incubator for 30–45 min to allow for complete gelation of the tissues.
18. While tissues are gelling, gather enough 12-well non-treated plates to accommodate one double-myobundle per well.
19. Warm 3D growth media to 37 °C and add 1.5 mL of media per well.
20. After the tissues have fully gelled, the frame (with myobundles attached) can be removed from the mold for further culture.
21. To remove tissues from the molds, hold and stabilize the sides of the mold using a pair of forceps. Use another pair of forceps to remove the engineered tissue from the mold by gently sliding the fine-point forceps under the Velcro® frame and lifting the frame upward. Continue gently lifting the frame at several points along one edge of the mold. When the center of the hydrogel (the area not supported by the frame) has separated from the PDMS mold, repeat the lifting technique on each of the remaining three sides until entire tissue is separated. Only touch the frame with the forceps—touching the hydrogel with forceps will damage the tissue
22. Gently transfer the frames with double-myobundles to a media-containing well of the 12-well plate.
23. Repeat steps 21–22 until all tissues are removed from their molds (*see* **Note 16**).
24. Secure the lid of the 12-well plate(s) with tape.
25. Secure the plate(s) to a high-angle rocker in a 37°C, 5% CO_2 incubator. Tissues should be cultured free-floating in the culture media (Fig. 1).

3.7 Differentiation of Myobundles into Functional Skeletal Muscle

All steps should be performed in a BSL-2 biosafety cabinet. All myobundle incubation steps should be performed in a standard cell culture incubator set to 37°C, 5% CO_2 with cell culture plates secured on a high-angle rocker (Fig. 3A).

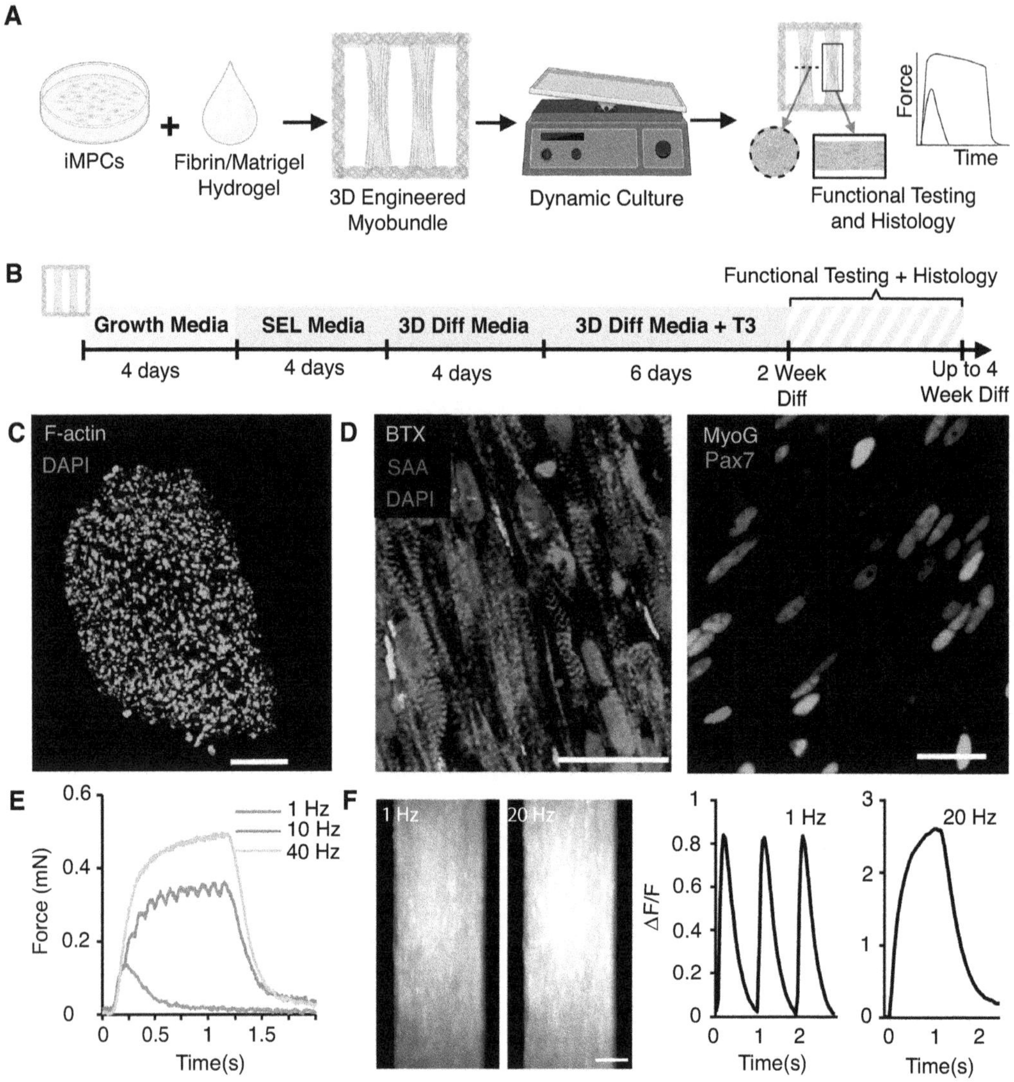

Fig. 3 iSKM myobundles are cross-striated and functional with pools of Pax7+ satellite cells. (**A**) Schematic of iSKM myobundle formation. (**B**) Standard myobundle culture timeline with growth period in green and differentiation period in blue. (**C**) Representative immunostaining of a cross-section from 2-week myobundles shows uniform and dense myotubes labeled with F-actin. Scale bar = 75 μm. (**D**) Representative longitudinal sections of 2-week myobundles show sarcomeric alpha actin (SAA) positive cross-striated myotubes with α-bungarotoxin (BTX)-labeled acetylcholine receptors (left) as well as pools of Pax7+ nuclei among MyoG+ myonuclei (right). Scale bars = 25 μm. (**E**) Representative force traces of a 4-week myobundle show increasing contractile force with increasing stimulation frequency and formation of tetanic contraction. (**F**) Representative peak GCaMP6 fluorescence images (left) and corresponding traces (right) during electrically stimulated calcium transients in 2-week myobundles. Scale bar = 100 μm. (Reproduced from [23] with permission.)

iSKM myobundle differentiation includes a 4-day growth period followed by a 2- to 4-week differentiation period (Fig. 3B). SEL media, 3D differentiation media, and 3D differentiation media supplemented with T3 are applied successively during the differentiation period (Fig. 3B).

1. Day 0–2: After 2 days of culture, replace the 3D growth media in each well, careful not to aspirate or otherwise disturb the myobundles.
2. Day 4–8: On day 4, change to SEL media and replace with fresh SEL media on day 6.
3. Day 8–12: On day 8, change to 3D differentiation media and replace with fresh 3D differentiation media every other day.
4. Day 12–18: On day 12, change to 3D differentiation media supplemented with 5 nM T3 and replate with fresh 3D differentiation media supplemented with 5 nM T3 every other day. Day 18 marks the completion of a 2-week differentiation period, but this step 5 feeding regimen can be continued out to 4 weeks (day 32) if needed.
5. Analyze the tissues for structure (Fig. 3C, D), force generation (Fig. 3E), and electrical propagation (Fig. 3F) 2–4 weeks after initial tissue formation.

4 Notes

1. We have also used Cerex® nylon as a frame material, finding it tends to absorb more of the cell/gel solution, resulting in smaller final tissues. However, Cerex® nylon can be used with a slightly higher volume of cell/gel solution if there are difficulties manufacturing Velcro® frames. If using Velcro®, be sure to use the soft "loop" side rather than the rough "hook" side for making frames.
2. Our plasmid (Dox-inducible Pax7 with GFP reporter) is available on Addgene and listed as Plasmid #109372. M2rtTA transactivator plasmid FUW-M2rtTA, packaging plasmid psPAX2, and envelope plasmid pMD2.G are listed as Plasmids #20342, #12260, and #12259, respectively.
3. We have also maintained hiPSCs using E8 medium as alternative to mTESR Plus.
4. Cells are sorted for GFP by staff at the Duke University Flow Cytometry Core using the MoFlo Astrios Cell Sorter with a 70 μm nozzle.
5. PDMS molds can be generated weeks to months in advance and stored in 70% ethanol solution until use. PDMS molds

can be reused after tissues are removed but must be sterilized for reuse via step 9 in Sect. 3.1.

6. To EDTA passage confluent hiPSCs, rinse with 1× DPBS and apply 0.5 mM EDTA. Incubate for ~3 min at room temperature before aspirating off the EDTA. Using a P1000 micropipette, use mTESR™ media to pipette the colonies off of the well, pipetting no more than three times total. Plate cell suspension in mTESR™ supplemented with 5 μM Y-27632 at a 1:100 to 1:40 (vol/vol) ratio. Rinse with 1× DPBS the following day before maintaining in mTESR™.
7. To coat tissue culture plates/flasks with Matrigel®, add 10 μL Matrigel®/mL cold 1× DPBS. Add enough Matrigel® solution to coat the bottom of the culture vessel and incubate at 37°C for at least 30 min. Aspirate solution from the culture vessel immediately before use.
8. To test transduction efficiency, which can vary based on virus titer and cell line, we recommend testing 1, 2, 4, and 6 μL of each virus per mL of culture media with 4 μg/mL polybrene in an overnight incubation. Several days after transduction, apply DOX at 1 μg/mL of culture media. Measure GFP expression and cell viability 24–48 h after DOX administration to determine optimal transduction protocol.
9. Transduced hiPSCs can be frozen in CryoStor® CS10 during expansion and stored long term in liquid nitrogen for future differentiations.
10. Cells will begin to grow on top of each other, but do not passage. At time of sort, there should be dark colonies of cells on top of a "feeder layer" of cells and ECM. Cells can be sorted starting at day 19 and out to day 23, maintaining in E6 media supplemented with DOX and bFGF.
11. If cells are still heavily clumped, cell suspension can be passed through a sterile syringe with increasing needle gauges (decreasing diameters). We have used 18 G, 21 G, and then 23 G needles in this application before moving the cell suspension through a 100-μm cell strainer, followed by the 40-μm cell strainer.
12. These cells can only be passaged one time before being plated for myobundle fabrication. iMPCs passaged at this step can also be cryopreserved in freezing media containing 90% FBS and 10% dimethylsulfoxide (vol/vol).
13. Fibrinogen solution should be prepared less than 4 h before use and kept on ice. Sterile filtration should occur when the solution is still warm, and then the solution should be placed on ice. Ensure that the solution is fully cooled to 4°C before use.

14. For our molds accommodating two (5 mm long, 1 mm diameter) myobundles, the total volume needed to fill the mold is 38 μL (19 μL per myobundle). Thus, with a 1:1 ratio of gel and cell solutions, we combine 20 μL gel solution with 20 μL cell solution to fill one mold, accounting for minimal volume loss during pipetting.
15. If the hydrogel begins to take on a viscous appearance while being injected into the molds, the hydrogel has begun to polymerize. The number of tissues made at one time (step 10, Sect. 3.6) should be decreased.
16. If the tissue is adhering to the PDMS mold, the Pluronic F-127 solution can be used at a higher concentration or applied to the molds for a longer period of time.

Acknowledgments

This work was supported by the National Institutes of Health (NIH) grants UG3TR002142 (N.B.), U01EB028901 (N.B.), R01AR070543 (N.B.), R01AR079223 (N.B.), R01ARO82979 (N.B., A.K.), R21AR078269 (NB), Jain Foundation (N.B.), and National Science Foundation (NSF) Graduate Research Fellowship (A.D.).

References

1. Campbell KP (1995) Three muscular dystrophies: loss of cytoskeleton-extracellular matrix linkage. Cell 80(5):675–679. https://doi.org/10.1016/0092-8674(95)90344-5
2. Rayavarapu S, Coley W, Kinder TB, Nagaraju K (2013) Idiopathic inflammatory myopathies: pathogenic mechanisms of muscle weakness. Skelet Muscle 3(1):13. https://doi.org/10.1186/2044-5040-3-13
3. Powers SK, Lynch GS, Murphy KT, Reid MB, Zijdewind I (2016) Disease-induced skeletal muscle atrophy and fatigue. Med Sci Sports Exerc 48(11):2307–2319. https://doi.org/10.1249/MSS.0000000000000975
4. Wells DJ (2018) Tracking progress: an update on animal models for Duchenne muscular dystrophy. Dis Model Mech 11(6). https://doi.org/10.1242/dmm.035774
5. Chiu YH, Hornsey MA, Klinge L, Jorgensen LH, Laval SH, Charlton R, Barresi R, Straub V, Lochmuller H, Bushby K (2009) Attenuated muscle regeneration is a key factor in dysferlin-deficient muscular dystrophy. Hum Mol Genet 18(11):1976–1989. https://doi.org/10.1093/hmg/ddp121
6. Khodabukus A (2021) Tissue-engineered skeletal muscle models to study muscle function, plasticity, and disease. Front Physiol 12:619710. https://doi.org/10.3389/fphys.2021.619710
7. Chen W, Nyasha MR, Koide M, Tsuchiya M, Suzuki N, Hagiwara Y, Aoki M, Kanzaki M (2019) In vitro exercise model using contractile human and mouse hybrid myotubes. Sci Rep 9(1):11914. https://doi.org/10.1038/s41598-019-48316-9
8. Eberli D, Soker S, Atala A, Yoo JJ (2009) Optimization of human skeletal muscle precursor cell culture and myofiber formation in vitro. Methods 47(2):98–103. https://doi.org/10.1016/j.ymeth.2008.10.016
9. Khodabukus A, Prabhu N, Wang J, Bursac N (2018) In Vitro tissue-engineered skeletal muscle models for studying muscle physiology and disease. Adv Healthc Mater 7(15):e1701498. https://doi.org/10.1002/adhm.201701498
10. Philippi S, Bigot A, Marg A, Mouly V, Spuler S, Zacharias U (2012) Dysferlin-deficient immortalized human myoblasts and myotubes as a useful tool to study dysferlinopathy. PLoS Curr

4:RRN1298. https://doi.org/10.1371/currents.RRN1298
11. Machida S, Spangenburg EE, Booth FW (2004) Primary rat muscle progenitor cells have decreased proliferation and myotube formation during passages. Cell Prolif 37(4):267–277. https://doi.org/10.1111/j.1365-2184.2004.00311.x
12. Cosgrove BD, Sacco A, Gilbert PM, Blau HM (2009) A home away from home: challenges and opportunities in engineering in vitro muscle satellite cell niches. Differentiation 78(2-3):185–194. https://doi.org/10.1016/j.diff.2009.08.004
13. Voloshin N, Tyurin-Kuzmin P, Karagyaur M, Akopyan Z, Kulebyakin K (2023) Practical use of immortalized cells in medicine: current advances and future perspectives. Int J Mol Sci 24(16). https://doi.org/10.3390/ijms241612716
14. Cardone N, Taglietti V, Baratto S, Kefi K, Periou B, Gitiaux C, Barnerias C, Lafuste P, Pharm FL, Pharm JN, Panicucci C, Desguerre I, Bruno C, Authier FJ, Fiorillo C, Relaix F, Malfatti E (2023) Myopathologic trajectory in Duchenne muscular dystrophy (DMD) reveals lack of regeneration due to senescence in satellite cells. Acta Neuropathol Commun 11(1):167. https://doi.org/10.1186/s40478-023-01657-z
15. Takahashi K, Tanabe K, Ohnuki M, Narita M, Ichisaka T, Tomoda K, Yamanaka S (2007) Induction of pluripotent stem cells from adult human fibroblasts by defined factors. Cell 131(5):861–872. https://doi.org/10.1016/j.cell.2007.11.019
16. Yu J, Vodyanik MA, Smuga-Otto K, Antosiewicz-Bourget J, Frane JL, Tian S, Nie J, Jonsdottir GA, Ruotti V, Stewart R, Slukvin II, Thomson JA (2007) Induced pluripotent stem cell lines derived from human somatic cells. Science 318(5858):1917–1920. https://doi.org/10.1126/science.1151526
17. Caron L, Kher D, Lee KL, McKernan R, Dumevska B, Hidalgo A, Li J, Yang H, Main H, Ferri G, Petek LM, Poellinger L, Miller DG, Gabellini D, Schmidt U (2016) A human pluripotent stem cell model of facioscapulohumeral muscular dystrophy-affected skeletal muscles. Stem Cells Transl Med 5(9):1145–1161. https://doi.org/10.5966/sctm.2015-0224
18. Tanaka A, Woltjen K, Miyake K, Hotta A, Ikeya M, Yamamoto T, Nishino T, Shoji E, Sehara-Fujisawa A, Manabe Y, Fujii N, Hanaoka K, Era T, Yamashita S, Isobe K, Kimura E, Sakurai H (2013) Efficient and reproducible myogenic differentiation from human iPS cells: prospects for modeling Miyoshi Myopathy in vitro. PLoS One 8(4):e61540. https://doi.org/10.1371/journal.pone.0061540
19. Nesmith AP, Wagner MA, Pasqualini FS, O'Connor BB, Pincus MJ, August PR, Parker KK (2016) A human in vitro model of Duchenne muscular dystrophy muscle formation and contractility. J Cell Biol 215(1):47–56. https://doi.org/10.1083/jcb.201603111
20. Chal J, Al Tanoury Z, Hestin M, Gobert B, Aivio S, Hick A, Cherrier T, Nesmith AP, Parker KK, Pourquie O (2016) Generation of human muscle fibers and satellite-like cells from human pluripotent stem cells in vitro. Nat Protoc 11(10):1833–1850. https://doi.org/10.1038/nprot.2016.110
21. Abujarour R, Bennett M, Valamehr B, Lee TT, Robinson M, Robbins D, Le T, Lai K, Flynn P (2014) Myogenic differentiation of muscular dystrophy-specific induced pluripotent stem cells for use in drug discovery. Stem Cells Transl Med 3(2):149–160. https://doi.org/10.5966/sctm.2013-0095
22. Shoji E, Sakurai H, Nishino T, Nakahata T, Heike T, Awaya T, Fujii N, Manabe Y, Matsuo M, Sehara-Fujisawa A (2015) Early pathogenesis of Duchenne muscular dystrophy modelled in patient-derived human induced pluripotent stem cells. Sci Rep 5: 12831. https://doi.org/10.1038/srep12831
23. Rao L, Qian Y, Khodabukus A, Ribar T, Bursac N (2018) Engineering human pluripotent stem cells into a functional skeletal muscle tissue. Nat Commun 9(1). https://doi.org/10.1038/s41467-017-02636-4
24. Darabi R, Arpke RW, Irion S, Dimos JT, Grskovic M, Kyba M, Perlingeiro RC (2012) Human ES- and iPS-derived myogenic progenitors restore DYSTROPHIN and improve contractility upon transplantation in dystrophic mice. Cell Stem Cell 10(5):610–619. https://doi.org/10.1016/j.stem.2012.02.015
25. Tedesco FS, Gerli MF, Perani L, Benedetti S, Ungaro F, Cassano M, Antonini S, Tagliafico E, Artusi V, Longa E, Tonlorenzi R, Ragazzi M, Calderazzi G, Hoshiya H, Cappellari O, Mora M, Schoser B, Schneiderat P, Oshimura M, Bottinelli R, Sampaolesi M, Torrente Y, Broccoli V, Cossu G (2012) Transplantation of genetically corrected human iPSC-derived progenitors in mice with limb-girdle muscular dystrophy. Sci Transl Med 4(140):140ra189. https://doi.org/10.1126/scitranslmed.3003541
26. Goudenege S, Lebel C, Huot NB, Dufour C, Fujii I, Gekas J, Rousseau J, Tremblay JP (2012) Myoblasts derived from normal hESCs and dystrophic hiPSCs efficiently fuse with existing muscle fibers following transplantation. Mol Ther 20(11):2153–2167. https://doi.org/10.1038/mt.2012.188

27. Choi IY, Lim H, Estrellas K, Mula J, Cohen TV, Zhang Y, Donnelly CJ, Richard JP, Kim YJ, Kim H, Kazuki Y, Oshimura M, Li HL, Hotta A, Rothstein J, Maragakis N, Wagner KR, Lee G (2016) Concordant but varied phenotypes among Duchenne muscular dystrophy patient-specific myoblasts derived using a human iPSC-based model. Cell Rep 15(10):2301–2312. https://doi.org/10.1016/j.celrep.2016.05.016
28. Wang J, Khodabukus A, Rao L, Vandusen K, Abutaleb N, Bursac N (2019) Engineered skeletal muscles for disease modeling and drug discovery. Biomaterials 221:119416. https://doi.org/10.1016/j.biomaterials.2019.119416
29. Madden L, Juhas M, Kraus WE, Truskey GA, Bursac N (2015) Bioengineered human myobundles mimic clinical responses of skeletal muscle to drugs. eLife 4. https://doi.org/10.7554/eLife.04885
30. Wang J, Zhou CJ, Khodabukus A, Tran S, Han SO, Carlson AL, Madden L, Kishnani PS, Koeberl DD, Bursac N (2021) Three-dimensional tissue-engineered human skeletal muscle model of Pompe disease. Commun Biol 4(1):524. https://doi.org/10.1038/s42003-021-02059-4
31. Kondash ME, Ananthakumar A, Khodabukus A, Bursac N, Truskey GA (2020) Glucose uptake and insulin response in tissue-engineered human skeletal muscle. Tissue Eng Regen Med 17(6):801–813. https://doi.org/10.1007/s13770-020-00242-y
32. Khodabukus A, Kaza A, Wang J, Prabhu N, Goldstein R, Vaidya VS, Bursac N (2020) Tissue engineered human myobundle system as a platform for evaluation of skeletal muscle injury biomarkers. Toxicol Sci 176(1):124–136. https://doi.org/10.1093/toxsci/kfaa049
33. Chen Z, Li B, Zhan RZ, Rao L, Bursac N (2021) Exercise mimetics and JAK inhibition attenuate IFN-gamma-induced wasting in engineered human skeletal muscle. Sci Adv 7(4). https://doi.org/10.1126/sciadv.abd9502
34. Khodabukus A, Madden L, Prabhu NK, Koves TR, Jackman CP, Muoio DM, Bursac N (2019) Electrical stimulation increases hypertrophy and metabolic flux in tissue-engineered human skeletal muscle. Biomaterials 198:259–269. https://doi.org/10.1016/j.biomaterials.2018.08.058
35. Wang J, Broer T, Chavez T, Zhou CJ, Tran S, Xiang Y, Khodabukus A, Diao Y, Bursac N (2022) Myoblast deactivation within engineered human skeletal muscle creates a transcriptionally heterogeneous population of quiescent satellite-like cells. Biomaterials 284:121508. https://doi.org/10.1016/j.biomaterials.2022.121508
36. Broer T, Tsintolas N, Purkey K, Hammond S, DeLuca S, Wu T, Gupta I, Khodabukus A, Bursac N (2024) Engineered myovascular tissues for studies of endothelial/satellite cell interactions. Acta Biomater 188:65–78. https://doi.org/10.1016/j.actbio.2024.09.020
37. Khodabukus A, Prabhu NK, Roberts T, Buldo M, Detwiler A, Fralish ZD, Kondash ME, Truskey GA, Koves TR, Bursac N (2024) Bioengineered model of human LGMD2B skeletal muscle reveals roles of intracellular calcium overload in contractile and metabolic dysfunction in dysferlinopathy. Adv Sci (Weinh) 11(31):e2400188. https://doi.org/10.1002/advs.202400188
38. Helfer A, Bursac N (2021) Frame-hydrogel methodology for engineering highly functional cardiac tissue constructs. Methods Mol Biol 2158:171–186. https://doi.org/10.1007/978-1-0716-0668-1_13

Chapter 5

Bulk Differentiation of Human Induced Pluripotent Stem Cells into Contractile Skeletal Muscles

Yu-Shen Cheng, Farah Ulfat Jamee, Muhammad Irfanur Rashid, Hidetoshi Sakurai, Kazunori Shimizu and Yohei Okada

Abstract

Effective treatments for various muscular disorders and a clear understanding of their disease pathology remain lacking owing to the limitations of traditional animal models and two-dimensional (2D) cell cultures. Human induced pluripotent stem cells (hiPSCs) offer promising platforms for disease modeling and regenerative medicine. However, existing protocols for skeletal muscle differentiation are often labor-intensive and yield inconsistent results. Here, we utilized a piggyBac transposon-based system to generate bulk *MYOD1*-hiPSCs, eliminating the need for clonal selection and enabling rapid, efficient, and reproducible skeletal muscle differentiation. Furthermore, we established a protocol for generating functional three-dimensional (3D) skeletal muscle tissues that exert contractile force in response to electrical stimulation, highlighting their physiological relevance. This system provides a simple and valuable tool for disease modeling, drug screening, and the study of sporadic muscle disorders using patient-derived hiPSCs.

Keywords *MYOD1*-hiPSC, Bulk culture, Skeletal muscle differentiation, 3D culture, Contraction, Disease modeling

1 Introduction

Skeletal muscle is one of the most functionally diverse tissues in the human body, with essential roles in respiration, posture, and locomotion [1]. Skeletal muscle gradually loses functions owing to aging [2], muscular diseases [3], or volumetric skeletal muscle loss [4], resulting in debilitating myopathies. However, many skeletal muscle diseases still lack definitive treatments. In research on skeletal muscle diseases, traditional small animal models have been used to examine disease progression and identify pharmacological compounds [5]. However, mouse models often fail to fully recapitulate the pathological features of human muscular diseases. Therefore, the development of more accurate skeletal muscle models is essential to better predict clinically relevant treatment effects [6].

Joe Quadrilatero (ed.), *Myogenesis and Skeletal Muscle Biology*, Methods in Molecular Biology, vol. 3057,
https://doi.org/10.1007/978-1-0716-5396-8_5,

Human pluripotent stem cells (hPSCs), such as human embryonic stem cells (hESCs) and human induced pluripotent stem cells (hiPSCs), can differentiate into any type of somatic cell and hold great potential for clinical and research applications [7, 8]. Patient-derived hPSCs can differentiate into the affected cell type, offering more accurate disease models than traditional systems such as cell lines or animal models [9]. However, there are still some challenges in the differentiation of hPSCs into skeletal muscle, including the need to achieve rapid, efficient, and reproducible differentiation for disease modeling.

In recent decades, several differentiation protocols for generating skeletal muscle from hPSCs have been reported [10]. Some of these protocols differentiate hPSCs into skeletal muscle in vitro using various morphogens and small molecules, aiming to recapitulate the in vivo differentiation process and thereby achieve more physiologically relevant characteristics. However, these protocols often involve many steps and require significant time and labor [11–16]. In contrast, the overexpression of transcription factors involved in skeletal muscle development can more directly and rapidly induce hPSCs to differentiate into skeletal muscle. For example, hPSCs that overexpress *PAX3* and *PAX7*, which are expressed in satellite cells, can be used for skeletal muscle differentiation [17]. Additionally, *MYOD1*, a critical gene for myogenic differentiation that induces myogenic transdifferentiation from fibroblasts [18], is commonly used for myogenic differentiation from hPSCs and for the pathophysiological analysis of muscular diseases using patient-derived hPSCs [19–24]. Although these transgene-based protocols may bypass certain developmental processes and therefore may not fully mimic physiological development, they are still valuable tools for disease modeling and regenerative therapies because of their high efficiency [25].

To date, skeletal muscle differentiation via *MYOD1* overexpression has been achieved by using lentiviruses, adenoviruses, and piggyBac transposon vectors, and these methods are widely used in muscular disease research [10, 20–24, 26, 27]. The ability of the piggyBac vector system to induce the *MYOD1* expression in hPSCs allows highly efficient myogenic differentiation [22, 28]. However, time- and labor-intensive procedures are necessary to select an appropriate *MYOD1*-hPSC clone for optimal differentiation efficiency [29]. Additionally, clonal variation affects differentiation efficiency and phenotypic consistency, complicating disease modeling and drug screening.

Therefore, we established a bulk *MYOD1*-hiPSC method by transfecting hiPSCs with a piggyBac vector followed by selection with puromycin [30, 31]. This method allows highly efficient and rapid skeletal muscle differentiation without the need for clonal selection. Moreover, three-dimensional (3D) skeletal muscle tissues generated from bulk *MYOD1*-hiPSCs exhibited contractile force in response to electrical stimulation, indicating their functional ability.

The bulk *MYOD1*-hiPSC method enables rapid analysis of large hiPSC populations, making it especially useful for studying sporadic diseases using samples from patient registries.

Here, we introduce our protocol for two-dimensional (2D) skeletal muscle differentiation and the generation of functional 3D skeletal muscle tissue from bulk *MYOD1*-hiPSCs.

2 Materials

2.1 Human iPSC Culure and Skeletal Muscle Differentiation

1. Cell Culture Dish 90 mm.
2. 6- and 24-well culture plates.
3. DMEM/F12.
4. KnockOut™ Serum Replacement (KSR; Thermo Fisher Scientific, USA).
5. Non-essential amino acids.
6. 2-Mercaptoethanol.
7. Basic fibroblast growth factor (bFGF, FGF2; Pepro-Tech, USA).
8. iMatrix-511 silk (Nippi, Japan).
9. StemFit medium (AK02N, Ajinomoto, Japan).
10. Accutase (Nacalai Tesque, Japan).
11. Neon™ Transfection System (Thermo Fisher Scientific, USA).
12. Neon™ Transfection System Kit (10 μL) (Thermo Fisher Scientific, USA)
13. Y27632 (FUJIFILM Wako Pure Chemical Corporation, Japan).
14. Puromycin.
15. Matrigel.
16. Doxycycline (Dox).
17. αMEM.
18. Ultroser G (BioSera, Pall, USA).
19. High-glucose DMEM (with 1500 mg/L $NaHCO_3$).
20. Horse serum (HS).
21. Recombinant human insulin-like growth factor 1.
22. SB431542.
23. Microdevices made of polydimethylsiloxane (PDMS).
24. Pluronic F-127 solution.
25. Fibrinogen (10 mg/mL, freshly prepared).
26. Thrombin.
27. 6-Aminocaproic acid (6-AA).
28. trans-4-(Aminomethyl)cyclohexanecarboxylic acid (TA).

2.2 Measurement of Contractile Force

1. Electrical stimulator (C-Pace EP; IonOptix, USA).
2. Upright microscope.

2.3 Histological Analysis

1. Paraformaldehyde (PFA).
2. Phosphate-buffered saline (PBS).
3. Fetal bovine serum (FBS).
4. Triton X-100.
5. Primary antibodies and dilution:
   - MyoD, 1:500.
   - HA-Tag, 1:1000.
   - MHC, 1:500.
   - MyoG, 1:500.
   - Oct3/4 1:100.
   - Nanog, 1:100.
   - Titin, 1:500.
   - α-Actinin, 1:500.
6. Alexa Fluor 488-, 555-, and 647-conjugated secondary antibodies.
7. Hoechst 33258.
8. Confocal laser scanning microscope

3 Methods

3.1 Human iPSC Culture Prior to MYOD1-hiPSC Generation

1. Cultivate hiPSCs on mitomycin C-treated SNL murine fibroblast feeder cells seeded on 0.1% gelatin-coated tissue culture dishes with hESC medium composed of DMEM/F12 supplemented with 20% KSR, non-essential amino acids, 0.1 mM 2-ME, and 4 ng/mL bFGF at 37°C in a humidified atmosphere containing 3% CO_2.
2. For feeder-free culture, transfer hiPSCs from on-feeder culture onto iMatrix-511 silk-coated plates and culture them in StemFit medium (AK02N) under a humidified atmosphere containing 5% CO_2.

3.2 Generation of MYOD1-hiPSCs

1. Dissociate semiconfluent hiPSCs under feeder-free conditions into single cells using Accutase.
2. Prepare the Neon transfection system and set the electroporation parameters to a pulse voltage of 1100 V, a pulse width of 30 ms, and one pulse.
3. Electroporate 1.25 × 10^5 cells with 0.5 μg of PB-TA-3HA-h*MYOD1*-ERP2 [30] and the transposase expression vector

CAG-PBase [32] using a 10 μL Neon tip (Fig. 1a). Be careful not to include bubbles in the tip during electroporation.

4. Plate the cells in a six-well plate with StemFit medium (AK02N) supplemented with 10 μM Y27632.
5. After 2 days, replace the medium with fresh StemFit medium containing 0.5 μg/mL puromycin.
6. Cultivate the *MYOD1*-hiPSCs for approximately 5–7 days until colony formation. Replace the medium with fresh puromycin-containing StemFit medium every 1–2 days.
7. Bulk-passage the puromycin-selected *MYOD1*-hiPSC colonies using Accutase and expand the cells in StemFit medium supplemented with 0.5 μg/mL puromycin to prepare frozen stocks (Fig. 1b, c).

3.3 2D Skeletal Muscle Differentiation from MYOD1-hiPSCs (Fig. 2)

1. Prepare Matrigel-coated plates using growth factor-reduced Matrigel at a dilution of 1:50–100 and incubate at 37°C for at least 2 h (*see* **Notes** 1).
2. Dissociate feeder-free *MYOD1*-hiPSCs into single cells by Accutase.

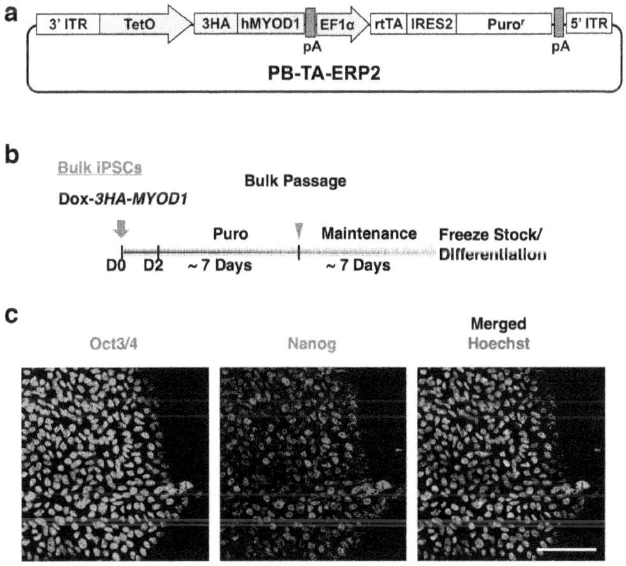

Fig. 1 Generation of bulk *MYOD1*-hiPSCs transduced with Dox-inducible 3HA-*hMYOD1* (Dox-3HA-MYOD1). (**a**) Construction of the piggyBac vector for Dox-3HA-*MYOD1* with a puromycin selection cassette. (**b**) Schematic presentation of the generation of bulk *MYOD1*-hiPSCs. (**c**) Immunocytochemistry (ICC) analysis of generated *MYOD1*-hiPSCs expressing the pluripotency markers (Oct3/4 (green) and Nanog (red)). Detect nuclei by Hoechst 33258 staining. All the *MYOD1*-hiPSCs express pluripotent markers. Scale bar, 50 μm. (Modified from [30, 31])

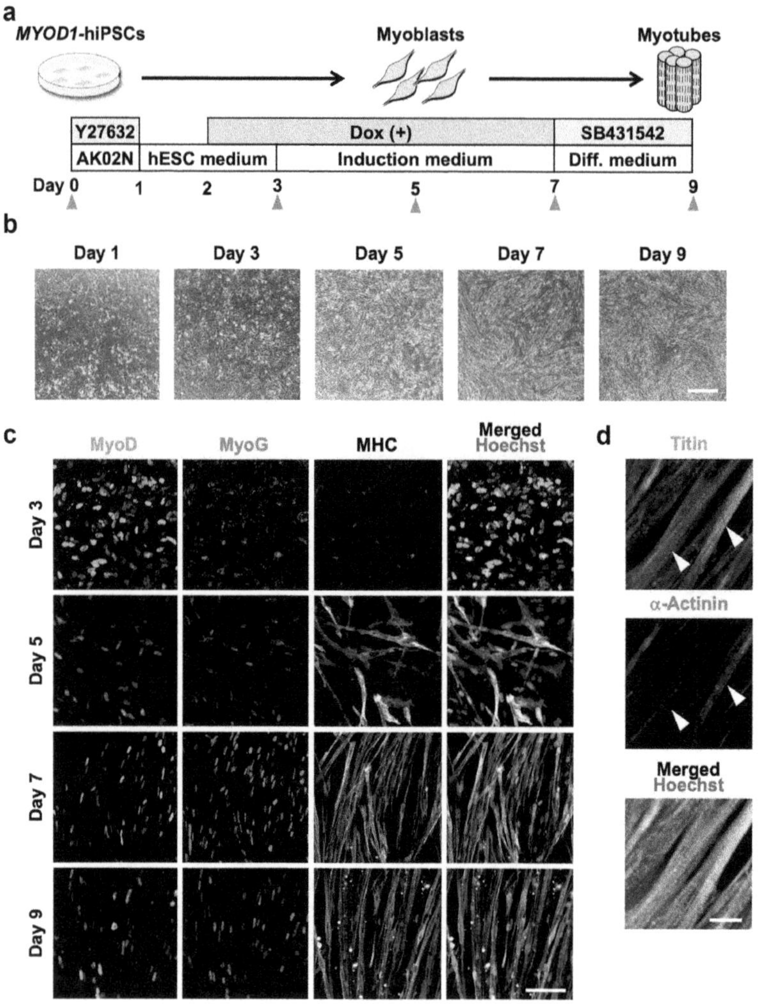

Fig. 2 Bulk *MYOD1*-hiPSCs successfully differentiated into mature myotubes. (**a**) Schematic presentation of the differentiation of *MYOD1*-hiPSCs into skeletal muscle. (**b**) Brightfield images showing the time course of myogenic differentiation of bulk *MYOD1*-hiPSCs. Scale bar, 100 μm. (**c, d**) Immunocytochemical analysis of the differentiating skeletal muscle for MyoD (green), myogenin (MyoG, red), MHC (white), and nuclei (blue). (**d**) Immunocytochemical analysis of the differentiating skeletal muscle for skeletal muscle markers, titin (green), α-actinin (red), and nuclei (blue). The arrowheads indicate sarcomere formation. Scale bar in (**c, d**), 200 μm. (Modified from [30, 31])

3. Plate 7×10^4 cells/well into Matrigel-coated 24-well plates or 1.3×10^4 cells/well into Matrigel-coated 96-well plates in StemFit medium (AK02N) containing 10 μM Y27632. The optimal cell density should be determined in advance depending on the hiPSC cell line and plate format used (*see* **Notes 2–4**).
4. On day 1, replace the medium with hESC medium without bFGF.

5. On day 2, replace the medium with fresh hESC medium without bFGF but supplemented with 1.5 μg/mL doxycycline to induce the expression of 3HA-h*MYOD1* (*see* **Note 5**).
6. On day 3, replace the medium with skeletal muscle induction medium consisting of α-MEM supplemented with 10% KSR, 2% Ultroser G, 0.1 mM 2-ME, and 1.5 μg/mL Dox (*see* **Note 6**) (Fig. 2b). MyoD and myogenin (MyoG) expression can be detected on day 3 (Fig. 2c).
7. On day 6 or 7, to promote myotube maturation, replace the medium with myotube differentiation medium composed of high-glucose DMEM (with 1500 mg/L $NaHCO_3$), 5% HS, 10 ng/mL IGF-1, and 10 μM SB431542. Most cells expressed skeletal muscle markers such as myosin heavy chain (MHC), α-actinin, and titin, and will form aligned myotube-like structures (Fig. 2b–d).
8. For long-term myotube culture, replace the medium with fresh myotube differentiation medium every 2 days.

3.4 Fabrication of 3D Myotubes from MYOD1-hiPSCs (Fig. 3a)

1. Prepare microdevices made of PDMS and place them at the center of each well of a 48-well plate [33, 34].
2. Sterilize the plate with the microdevices under UV light for at least 1 h.
3. Coat the devices with 2% Pluronic F-127 solution at room temperature for 1 h or incubate overnight at 4°C to prevent hydrogel adhesion.
4. Dissociate semiconfluent feeder-free *MYOD1*-hiPSCs into single cells using Accutase.
5. Seed *MYOD1*-hiPSCs in a six-well plate at a density of 3.5×10^5 cells/well using the same protocol used for 2D skeletal muscle differentiation.
6. On day 2, replace the medium with hESC medium without bFGF but supplemented with 1.5 μg/mL Dox.
7. On day 3, dissociate the cells by Accutase and resuspend them in skeletal muscle induction medium.
8. Prepare a hydrogel mixture by combining ice-cold fibrinogen (10 mg/mL, freshly prepared), Matrigel, and 2× DMEM at a volume ratio of 0.2:0.1:0.2.
9. Mix the cells ($6–8 \times 10^5$ cells/mL) with the hydrogel at a volume ratio of 0.484:0.5.
10. Add 50 U/mL thrombin to the mixture at a volume ratio of 0.016:1, ensuring that the thrombin is added last.
11. Load 45 μL of the final mixture (containing ~1.3×10^5 cells) with cold tips into the dumbbell-shaped pocket of each

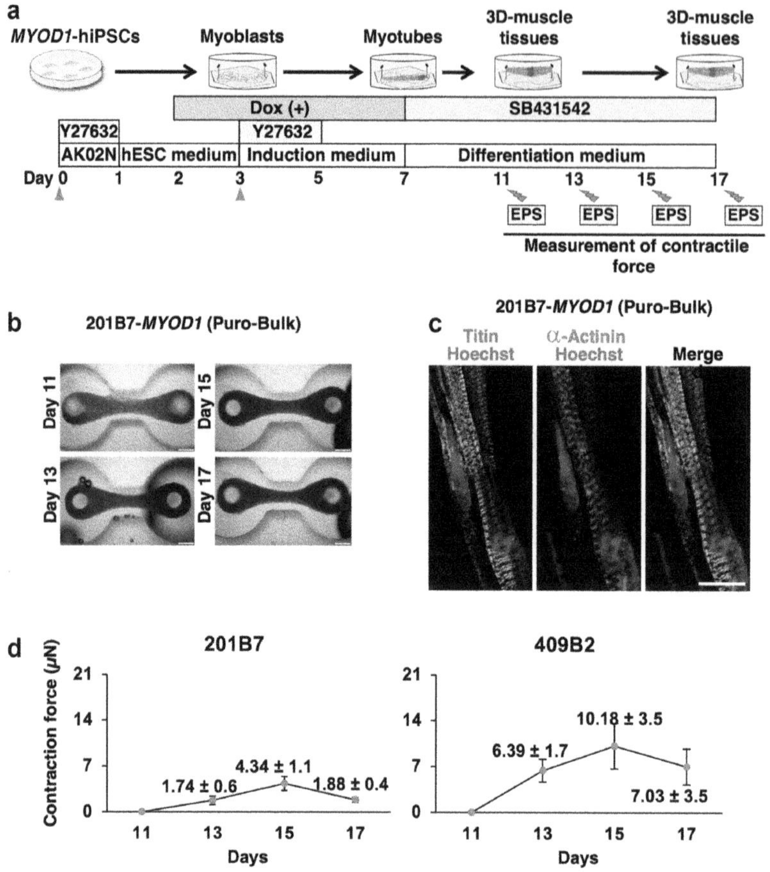

Fig. 3 The 3D myotubes fabrication from bulk *MYOD1*-hiPSCs exhibited contractile force. (**a**) Schematic presentation of the fabrication of 3D skeletal muscle tissue. On day 3 of differentiation, the cells are replated onto the microdevices. On day 11, 3D skeletal muscle tissue are pulled up at the top of the pillar, and the contractile force was measured in response to electrical pulse stimulation (EPS) on days 11, 13, 15, and 17. (**b**) Brightfield top-view images of 3D skeletal muscle tissue after 11, 13, 15, and 17 days of differentiation. Scale bar, 500 μ m. (**c**) IHC analysis of skeletal muscle markers, titin (green) and α-actinin (red), in 3D skeletal muscle tissue on day 17. Detect nuclei by Hoechst 33258 staining. Scale bar, 20 μm. (**d**) Measurement of the contractile force of 3D myotubes on day 11, 13, 15, and 17. (Modified from [30, 31])

microdevice and incubate at 37°C to allow the hydrogel to solidify, ensuring there are no bubbles.

12. Culture the cells in skeletal muscle induction medium supplemented with 0.5% penicillin/streptomycin, 1.5 μg/mL Dox, and 10 μM Y27632 for 2 days (*see* **Note 7**).
13. On day 5, replace the medium with the fresh induction medium without Y27632.
14. On day 7, replace the medium with differentiation medium (DMEM, 5% horse serum, 10 μM SB431542, and 0.5% penicillin/streptomycin, without Dox). Change medium every other day until day 17.

15. On day 11, the 3D skeletal muscle tissues are pulled up at the top of the pillar. By day 17, mature skeletal muscle with sarcomere formation will be observed (Fig. 3b, c). The skeletal muscle tissue exhibits contractile forces after 13 days of differentiation (Fig. 3d).
16. Throughout the entire culture period, include 2.0 mg/mL 6-AA and 1.0 mg/mL TA in all media to prevent tissue disassembly.

3.5 Measurement of Contractile Force

1. To electrically stimulate the skeletal muscle tissues to achieve maximum tetanic force, prepare an electrical stimulator with customized electrodes and set an electrical stimulus of 4.0 V/mm at 30 Hz with 2-ms-wide pulses.
2. Stimulate each tissue construct twice for 5 s on days 11, 13, 15, and 17 of differentiation.
3. Observe the displacement of the micropost tips during stimulation using an upright microscope equipped with a 4× (UPLFN4X) or a 20× objective (SLMPLN20X).
4. Capture images with and without stimulation.
5. Use ImageJ software to measure the displacement (δ) of each micropost tip and calculate the average displacement from both microposts in each tissue.
6. Calculate contractile force (F) using the following formula: $F = 3\pi ER^4\delta/(4L^3)$, where E is the elastic modulus of the PDMS (1.7 MN/m^2), R is the radius of a micropost (0.25 mm), δ is the displacement of the tip of the micropost, and L is the length of the micropost (4 mm) [33, 35].

3.6 Histological Analysis

3.6.1 2D Skeletal Muscles

1. Fix the cells with 4% PFA for 15 min at room temperature.
2. Wash the samples with PBS three times to remove the PFA.
3. Block and permeabilize the cells using blocking buffer consisting of PBS containing 10% FBS and 0.3% Triton X-100 for 1 h at room temperature.
4. Incubate the samples with primary antibodies overnight at 4°C (*see* **Note 8**).
5. Wash the samples with PBS three times to remove unbound primary antibodies.
6. Incubate the samples with Alexa Fluor 488-, 555-, or 647-conjugated secondary antibodies for 1 h at room temperature. Add 10 μg/mL Hoechst 33258 to stain the nuclei.

7. Observe the samples using a confocal laser scanning microscope (LSM700 or LSM900; Carl Zeiss, Germany). Randomly select and capture five images per sample.
8. Quantify the number of immunopositive cells and nuclei, and measure the immunopositive area and myotube thickness manually using ImageJ software [36].

3.6.2 3D Skeletal Muscle Tissue

1. Fix the tissues with 4% PFA for 30 min at room temperature.
2. Wash the samples with PBS three times to remove the PFA.
3. Permeabilize with PBS containing 0.3% Triton X-100 for 10 min at room temperature.
4. Wash the samples with PBS three times to remove the Triton X-100.
5. Block the samples with blocking buffer supplemented with PBS containing 10% goat serum and 0.01% Triton X-100 for 30 min at room temperature.
6. Incubate the samples with primary antibodies for 3 h at room temperature.
7. Wash the samples with PBS three times to remove unbound primary antibodies.
8. Incubate the samples with Alexa Fluor 488- or 555-conjugated secondary antibodies for 2 h at room temperature. Add 10 μg/mL Hoechst 33258 to stain the nuclei.
9. Observe the samples using a confocal laser scanning microscope and capture images.

4 Notes

1. Instead of coating culture plates with Matrigel prior to plating hiPSCs for skeletal muscle differentiation, the plates may be coated with recombinant laminin-511, laminin-421, or p421—a recombinant form of the D1-HS domain of perlecan conjugated to the C-terminus of laminin 421-E8 [37].
2. Y27632 can be removed from the hESC medium on day 1 of skeletal muscle differentiation. The morphology of the cell colonies may appear similar to that of iPSC-like colonies in the absence of Y27632 between days 1 and 2. However, the efficiency of skeletal muscle differentiation is comparable to that observed in the presence of Y27632.
3. The plating density of hiPSCs is critical for successful skeletal muscle differentiation. Low cell density can lead to poor

differentiation efficiency and cell death before maturation. In contrast, extremely high cell density may result in the formation of more iPSC-like colonies. Additionally, the optimal cell density varies depending on the hiPSC line and well size. Therefore, the cell density should be carefully optimized prior the main experiments [31].

4. Do not shake the 96-well plate after plating the hiPSC. Shaking may cause the cells to be distributed unevenly. Allow the plate to settle for a few minutes after plating so that the cells can attach to the bottom of the plate evenly before transferring the plate to the incubator.
5. Change the medium quickly and gently. Delays during medium changes can lead to cell drying. Moreover, adding or removing medium with too much force may cause cell detachment.
6. Optimize the concentration of Dox carefully. The Dox concentration significantly affects the efficiency of skeletal muscle differentiation. Accurate preparation of the desired concentration is essential.
7. SB431542 may be added to the induction medium from day 3 to day 7 during 3D skeletal muscle differentiation to increase the contractile force and promote myotube maturation [38].
8. Optimize the dilution of each primary antibody before the immunocytochemical experiments, as the optimal dilution varies with each antibody. Prepare and apply the correct dilution carefully to ensure reliable staining results.

Acknowledgments

The authors are grateful to Dr. S. Yamanaka (Kyoto University) for hiPSCs (201B7 and 409B2); Dr. K. Woltjen (Kyoto University) for the piggyBac vector for doxycycline-inducible expression system; the Institute of Comprehensive Medical Research, Division of Advanced Research Promotion, Aichi Medical University for their support in this project; Ms. M. Ishihara, Ms. Y. Kitashiba, Ms. S. Tokouda, Ms. E. Kojima, Ms. S. Yokoi, and Ms. Y. King for technical and administrative support; and all members of the Okada laboratory for assistance and discussions. This project was supported by the Japan Society for the Promotion of Science (JSPS) KAKENHI (grant numbers JP18K07539, JP19H03576, JP19K07969, JP21K07471, JP22H02988, JP23K06955, and JP25K02585), the Japan Agency for Medical Research and Development (AMED) (grant numbers JP19ek0109243, JP22bm0804020, and JP25bm1423003), and the Hori Sciences and Arts Foundation.

References

1. Janssen I, Heymsfield SB, Wang ZM, Ross R (2000) Skeletal muscle mass and distribution in 468 men and women aged 18-88 yr. J Appl Physiol (1985) 89(1):81–88. https://doi.org/10.1152/jappl.2000.89.1.81
2. Alway SE, Myers MJ, Mohamed JS (2014) Regulation of satellite cell function in sarcopenia. Front Aging Neurosci 6:246. https://doi.org/10.3389/fnagi.2014.00246
3. Wallace GQ, McNally EM (2009) Mechanisms of muscle degeneration, regeneration, and repair in the muscular dystrophies. Annu Rev Physiol 71:37–57. https://doi.org/10.1146/annurev.physiol.010908.163216
4. Grogan BF, Hsu JR, Skeletal Trauma Research C (2011) Volumetric muscle loss. J Am Acad Orthop Surg 19(Suppl 1):S35–37. https://doi.org/10.5435/00124635-201102001-00007
5. McGreevy JW, Hakim CH, McIntosh MA, Duan D (2015) Animal models of Duchenne muscular dystrophy: from basic mechanisms to gene therapy. Dis Model Mech 8(3):195–213. https://doi.org/10.1242/dmm.018424
6. Govoni A et al (2013) Ongoing therapeutic trials and outcome measures for Duchenne muscular dystrophy. Cell Mol Life Sci 70(23):4585–4602. https://doi.org/10.1007/s00018-013-1396-z
7. Thomson JA et al (1998) Embryonic stem cell lines derived from human blastocysts. Science 282(5391):1145–1147. https://doi.org/10.1126/science.282.5391.1145
8. Takahashi K et al. (2007) Induction of pluripotent stem cells from adult human fibroblasts by defined factors. Cell 131(5):861–872. https://doi.org/10.1016/j.cell.2007.11.019
9. Corti S, Faravelli I, Cardano M, Conti L (2015) Human pluripotent stem cells as tools for neurodegenerative and neurodevelopmental disease modeling and drug discovery. Expert Opin Drug Discov 10(6):615–629. https://doi.org/10.1517/17460441.2015.1037737
10. Chal J, Pourquie O (2017) Making muscle: skeletal myogenesis in vivo and in vitro. Development 144(12):2104–2122. https://doi.org/10.1242/dev.151035
11. Xu C et al (2013) A zebrafish embryo culture system defines factors that promote vertebrate myogenesis across species. Cell 155(4):909–921. https://doi.org/10.1016/j.cell.2013.10.023
12. Hosoyama T et al (2014) Derivation of myogenic progenitors directly from human pluripotent stem cells using a sphere-based culture. Stem Cells Transl Med 3(5):564–574. https://doi.org/10.5966/sctm.2013-0143
13. Borchin B, Chen J, Barberi T (2013) Derivation and FACS-mediated purification of PAX3+/PAX7+ skeletal muscle precursors from human pluripotent stem cells. Stem Cell Reports 1(6):620–631. https://doi.org/10.1016/j.stemcr.2013.10.007
14. Choi IY et al (2016) Concordant but varied phenotypes among Duchenne muscular dystrophy patient-specific myoblasts derived using a human iPSC-based model. Cell Rep 15(10):2301–2312. https://doi.org/10.1016/j.celrep.2016.05.016
15. Chal J et al (2016) Generation of human muscle fibers and satellite-like cells from human pluripotent stem cells in vitro. Nat Protoc 11(10):1833–1850. https://doi.org/10.1038/nprot.2016.110
16. van der Wal E et al (2018) Large-scale expansion of human iPSC-derived skeletal muscle cells for disease modeling and cell-based therapeutic strategies. Stem Cell Reports 10(6):1975–1990. https://doi.org/10.1016/j.stemcr.2018.04.002
17. Darabi R et al (2012) Human ES- and iPS-derived myogenic progenitors restore DYSTROPHIN and improve contractility upon transplantation in dystrophic mice. Cell Stem Cell 10(5):610–619. https://doi.org/10.1016/j.stem.2012.02.015
18. Davis RL, Weintraub H, Lassar AB (1987) Expression of a single transfected cDNA converts fibroblasts to myoblasts. Cell 51(6):987–1000. https://doi.org/10.1016/0092-8674(87)90585-x
19. Warren L et al (2010) Highly efficient reprogramming to pluripotency and directed differentiation of human cells with synthetic modified mRNA. Cell Stem Cell 7(5):618–630. https://doi.org/10.1016/j.stem.2010.08.012
20. Goudenege S et al (2012) Myoblasts derived from normal hESCs and dystrophic hiPSCs efficiently fuse with existing muscle fibers following transplantation. Mol Ther 20(11):2153–2167. https://doi.org/10.1038/mt.2012.188
21. Rao L et al (2012) Highly efficient derivation of skeletal myotubes from human embryonic stem cells. Stem Cell Rev Rep 8(4):1109–1119. https://doi.org/10.1007/s12015-012-9413-4
22. Tanaka A et al (2013) Efficient and reproducible myogenic differentiation from human iPS cells: prospects for modeling Miyoshi Myopathy in vitro. PLoS One 8(4):e61540. https://doi.org/10.1371/journal.pone.0061540
23. Tedesco FS et al (2012) Transplantation of genetically corrected human iPSC-derived progenitors in mice with

limb-girdle muscular dystrophy. Sci Transl Med 4(140):140ra189. https://doi.org/10.1126/scitranslmed.3003541

24. Lenzi J et al (2016) Differentiation of control and ALS mutant human iPSCs into functional skeletal muscle cells, a tool for the study of neuromuscolar diseases. Stem Cell Res 17(1):140–147. https://doi.org/10.1016/j.scr.2016.06.003
25. Shoji E, Woltjen K, Sakurai H (2016) Directed myogenic differentiation of human induced pluripotent stem cells. Methods Mol Biol 1353:89–99. https://doi.org/10.1007/7651_2015_257
26. Maffioletti SM et al (2015) Corrigendum: efficient derivation and inducible differentiation of expandable skeletal myogenic cells from human ES and patient-specific iPS cells. Nat Protoc 10(9):1457. https://doi.org/10.1038/nprot0915-1457d
27. Filareto A et al (2013) An ex vivo gene therapy approach to treat muscular dystrophy using inducible pluripotent stem cells. Nat Commun 4: 1549. https://doi.org/10.1038/ncomms2550
28. Uchimura T, Otomo J, Sato M, Sakurai H (2017) A human iPS cell myogenic differentiation system permitting high-throughput drug screening. Stem Cell Res 25:98–106. https://doi.org/10.1016/j.scr.2017.10.023
29. Yoshida T et al (2017) A skeletal muscle model of infantile-onset Pompe disease with patient specific iPS cells. Sci Rep 7(1):13473. https://doi.org/10.1038/s41598-017-14063-y
30. Rashid MI et al (2023) Simple and efficient differentiation of human iPSCs into contractible skeletal muscles for muscular disease modeling. Sci Rep 13(1):8146. https://doi.org/10.1038/s41598-023-34445-9
31. Okada Y (2024) Differentiation of human induced pluripotent stem cells (hiPSCs) into mature and contractible skeletal muscles. In: Ding B, Tang Y (eds) Human induced pluripotent stem cells. Springer, US, New York, NY, pp 147–161
32. Kim SI et al (2016) Inducible transgene expression in human iPS cells using versatile all-in-one piggyBac transposons. Methods Mol Biol 1357:111–131. https://doi.org/10.1007/7651_2015_251
33. Nagashima T et al (2020) In vitro model of human skeletal muscle tissues with contractility fabricated by immortalized human myogenic cells. Adv Biosyst 4(11):e2000121. https://doi.org/10.1002/adbi.202000121
34. Shimizu K et al (2017) Three-dimensional culture model of skeletal muscle tissue with atrophy induced by dexamethasone. Bioengineering (Basel) 4(2). https://doi.org/10.3390/bioengineering4020056
35. Vandenburgh H et al (2008) Drug-screening platform based on the contractility of tissue-engineered muscle. Muscle Nerve 37(4):438–447. https://doi.org/10.1002/mus.20931
36. Schneider CA, Rasband WS, Eliceiri KW (2012) NIH Image to ImageJ: 25 years of image analysis. Nat Methods 9(7):671–675. https://doi.org/10.1038/nmeth.2089
37. Zhao M et al (2024) Heparan sulfate chain-conjugated laminin-E8 fragments advance paraxial mesodermal differentiation followed by high myogenic induction from hiPSCs. Adv Sci (Weinh) 11(26):e2308306. https://doi.org/10.1002/advs.202308306
38. Nakamura S et al (2025) Construction of contractile human iPSC-derived skeletal muscle tissues on 96-well scale microdevices. J Biosci Bioeng 140:98–106. https://doi.org/10.1016/j.jbiosc.2025.05.003

Chapter 6

Quantification of C2C12 Myoblast Differentiation by Flow Cytometry

Göksu Oral, Rohan Chippalkatti and Daniel Kwaku Abankwa

Abstract

C2C12 myoblasts are widely used as an in vitro model to study skeletal muscle cell differentiation. Here, we update our previously published flow cytometry-based protocol to quantify C2C12 cell differentiation by measuring myosin heavy-chain (MyHC) marker expression. We explain the procedures for daily maintenance, transfection, induction of C2C12 cell differentiation, immunolabeling, and flow cytometry acquisition settings. We emphasize the importance of optimized cell seeding density to enhance transfection or differentiation efficiency and clarify the critical role of the correct cell population gate to accurately quantify MyHC-positive cells. Both the effect of drug treatments and genetic manipulations can be analyzed in this way. As an example, we quantify the disruption and rescue of differentiation only of the transformed, GFP-positive cell population after expression of mEGFP-tagged oncogenic K-Ras.

Keywords C2C12 myoblasts, Flow cytometry, K-Ras, Differentiation, Cancer, Rhabdomyosarcoma

1 Introduction

Cell differentiation is a fundamental process in metazoan life where cells gradually adopt their specialized phenotypes. This is not only critical throughout development but also for tissue homeostasis in the adult. Yet, in vitro models that accurately reproduce critical aspects of cell differentiation as seen in vivo are scarce [1].

The C2C12 skeletal muscle cell model is one of the best-established in vitro differentiation models, which recapitulates typical in vivo skeletal muscle differentiation steps [2]. We recently showed that the C2C12 cell line is heterogeneous [3]. It contains a small fraction of skeletal muscle stem cell-like progenitors, which are positive for the skeletal muscle stem cell transcription factor paired box 7 (Pax7) but negative for the late differentiation marker myosin II heavy chain (MyHC). In high serum (10% fetal bovine serum), progenitors proliferate and largely divide asymmetrically to replenish the major pool

The authors Göksu Oral and Rohan Chippalkatti contributed equally.

Joe Quadrilatero (ed.), *Myogenesis and Skeletal Muscle Biology*, Methods in Molecular Biology, vol. 3057, https://doi.org/10.1007/978-1-0716-5396-8_6,

of less proliferative, committed myoblastic Pax7−/MyHC− transit-amplifying cells. When the growth conditions are switched to low-serum medium (containing 2 % horse serum), transit-amplifying cells differentiate into Pax7−/ MyHC+ myocytes and myotubes.

We showed that oncogenic Ras mutants block terminal differentiation at the Pax7−/ MyHC− transit-amplifying cell stage. In the case of transformation with K-Ras4B-G12C, this phenotype could be cell-autonomously reversed by K-RasG12C inhibitors. Ras mutants were expressed as GFP-variant-tagged constructs, which could be specifically analyzed by gating for GFP-positive cells. We established a comprehensive flow cytometry-based analysis pipeline to enable manual and automated analysis of C2C12 cell differentiation [4]. Automation of gating steps was performed in a custom-written R script called FlowFate.

Here, we provide additional important instructions for our previous protocol [4], explaining the need for optimal side scatter (SSC) and forward scatter (FSC) settings for accurate quantification of differentiation. We furthermore elaborate on the impact of cell density on transfection and differentiation efficiencies.

2 Materials

2.1 Cell Culture

1. C2C12 mouse cell line (American Type Culture Collection (ATCC), cat. no.: CRL-1772).
2. Trypsin-EDTA (0.05 %) in phosphate-buffered saline (PBS).
3. The jetPRIME (Polyplus-transfection) transfection reagent for plasmid transfection.
4. High-serum medium: Dulbecco's modified Eagle's medium (DMEM) supplemented with ~9% (v/v) fetal bovine serum (FBS), 2 mM L-glutamine, and 1% (v/v) penicillin–streptomycin solution. Store at 4°C for up to 1 month.
5. Low-serum medium: DMEM supplemented with ~2% (v/v) horse serum (HS), 2 mM L-glutamine, and 1% (v/v) penicillin–streptomycin solution. Store at 4°C for up to 1 month.
6. AMG 510 (MedChemExpress, cat. no.: HY-11427), covalent K-RasG12C inhibitor [5] in dimethyl sulfoxide (DMSO) R99.5%.
7. The expression constructs used in this study include mEGFP-K-Ras4B and mEGFP-K-Ras4B-G12C, which were previously described [3]. Additionally, we utilized pEGFP-C2, obtained from Cytoskeleton Inc. (NovoPro, cat. no.: V012023).

2.2 MyHC Immunolabeling and Flow Cytometry

1. Myosin 4 monoclonal antibody (MF20), eFluor 660, eBioscience (Thermo Fisher Scientific, cat. no.: 50-6503-82).

2. PBS-T: Add 25 μL of Tween 20 to 50 mL PBS. Store at 4°C for up to 1 month.
3. 4% paraformaldehyde (PFA) fixation solution: Add 10 mL 16% (w/v) paraformaldehyde (PFA) to 30 mL PBS. Store at 4°C for up to 1 month.
4. 0.5% Triton X-100 in PBS: Add 250 μL of Triton X-100 to 50 mL PBS. Store at 4°C for up to 1 month.
5. ISOTON II diluent (Beckman Coulter, cat. no.: 8448011).
6. Guava easyCheck Beads (Cytek Biosciences, cat. no.: 4500-0025).

2.3 Equipment

1. Water bath at 37°C.
2. Incubator supplemented with 5% CO_2 at 37°C.
3. Z1 particle counter.
4. Microcentrifuge.
5. Guava easyCyte 6HT 2 L flow cytometer (Cytek Biosciences).
6. Cuvette for Coulter counter.
7. T75 flask.
8. Eppendorf tubes, 1.5 mL.
9. Falcon tubes, 15 mL.
10. Falcon tubes, 50 mL.
11. Six-well plate.
12. 96-well plate.

3 Methods

This protocol describes the cell seeding, transfection, and immunolabeling for flow cytometric acquisition and subsequent analysis of C2C12 cell differentiation using an example sample set. To determine the appropriate number and type of samples, *see* **Note 1**.

3.1 Cell Maintenance, Passaging, and Seeding

1. Propagate the cells in a T75 culture flask by refreshing the high-serum medium every 2 days under standard cell culture conditions (37°C, 5% CO_2). Once the cells reach 50%–60% confluency, subculture them into a new T75 flask by following the steps below.
2. Warm up the high-serum medium, PBS, and trypsin-EDTA in a water bath at 37°C for 5–10 min before starting the experiment.
3. Turn on the cell counter (Beckman Coulter Z1 Counter) and flush it twice with Milli-Q water and twice with ISOTON II Diluent before use.

4. Remove the high-serum medium from the T75 flask and wash the cells with 10 mL of PBS.
5. Discard the PBS and add 5 mL 0.05 % trypsin-EDTA.
6. Incubate the cells at 37°C until the cells detach completely from the flask, typically within 5–10 min.
7. Examine the cells under an inverted microscope to ensure the cell layer is fully dispersed.
8. Add 5 mL of high-serum medium to the flask to neutralize the trypsin, then gently pipette up and down to evenly resuspend the cells.
9. Transfer the cell suspension to a 15 mL Falcon tube and centrifuge at 200 × *g* for 3 min at room temperature to collect the cell pellet.
10. Remove the supernatant and resuspend the cell pellet in 1 mL fresh high-serum medium.
11. Mix 50 μL of the cell suspension with 10 mL of ISOTON II Diluent (1:200 dilution of cell suspension and ISOTON II Diluent) and determine the cell concentration using the Z1 particle counter.
12. Dispense the appropriate volume of the cell suspension to each well of the six-well plate with 2 mL of high-serum medium to obtain a final density of 50,000 cells per well for experiments including transfection of genetic constructs (Sect. 3.2) or 100,000 cells per well for differentiation assay (Sect. 3.3) on the next day.
13. To propagate the cells, seed 10^6 cells to a T75 flask and incubate at 37°C in 5% CO_2 until they reach the desired confluency.
14. Gently move the six-well plate in both forward-backward and side-to-side motions to evenly distribute the cells. Let it sit at room temperature for about 20 min before putting it in the incubator to prevent cell clumping in the center of the wells.

3.2 Cell Transfection

This optional section describes the cell seeding procedure for transfection with genetic constructs that are analyzed for their impact on C2C12 cell differentiation. If no transfection is needed, go to Sect. 3.3 for normal differentiation or Sect. 3.4 if drug treatment during differentiation is intended.

It is necessary to seed cells at an optimal density for transfection, since a too high seeding density can lead to reduced transfection efficiency (Fig. 1). In addition to the target samples (here transfected with mEGFP-K-Ras4B wt and mEGFP-K-Ras4B-G12C encoding plasmids), an mEGFP-only control sample (here transfected with pEGFP-C2) is required. Other GFP variants can be employed; however, it is imperative that all fluorescent proteins of target samples

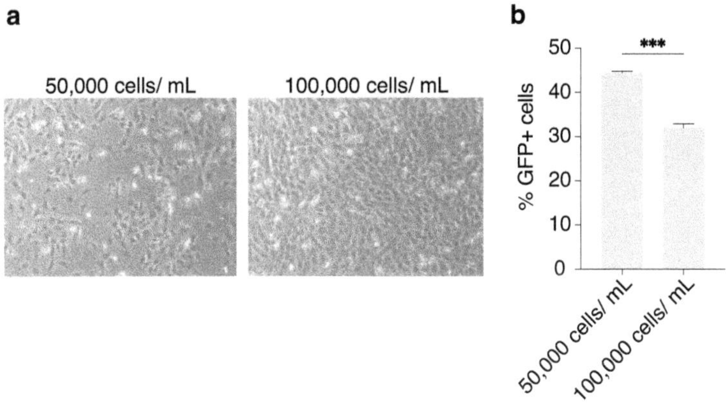

Fig. 1 The correct seeding density is required for a good transfection efficiency of C2C12 cells. (**a**) Brightfield microscopy images of C2C12 cells seeded at indicated densities in high-serum medium. Scale bar: 200 μm. (**b**) Transfection efficiency of an EGFP-control construct as the percentage of EGFP-positive C2C12 cells in dependence of the seeding density was determined using flow cytometric analysis. Data represent means ± SEM; $n = 3$ independent biological repeats. Statistical comparison was performed using the Mann–Whitney test. $^{***}p \leq 0.001$

and the control have the same spectroscopic properties, notably the brightness.

1. The cell confluency should be 50%–60% for transfection in a six-well plate.
2. Add 2 μg plasmid DNA to 200 μL jetPRIME buffer and mix by vortexing (*see* **Note 2**).
3. Vortex and spin down the jetPRIME reagent briefly before use.
4. Add 4 μL of jetPRIME reagent to the DNA-containing transfection solution, vortex, and centrifuge briefly.
5. Let the transfection mixture incubate for 10 min at 23–25°C.
6. Add 200 μL of transfection mixture dropwise into each well containing the 2 mL of high-serum medium and gently move the six-well plate back and forth as well as sideways to ensure uniform distribution.
7. 4 h after transfection, replace the medium with fresh high-serum medium (*see* **Note 3**).

3.3 Cell Differentiation

This section describes the transfection protocol of untreated or transfected C2C12 cells. If drug treatment is intended, go to Sect. 3.4.

C2C12 cell differentiation is sensitive to the cell confluency at the time of the switch from high- to low-serum medium. Cell confluency must be at least 90% to ensure optimal differentiation. Lower cell densities will reduce the extent of differentiation and can

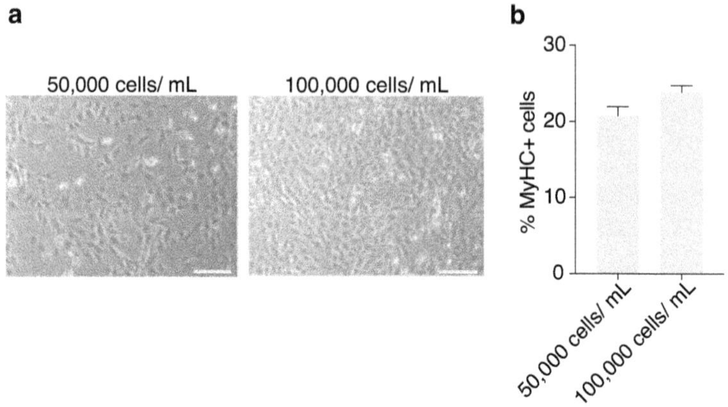

Fig. 2 Differentiation efficiency of C2C12 cells can depend on the seeding density. Differentiated cell fraction (% MyHC+ cells) was determined as described in Sect. 3.4. (**a**) Brightfield microscopy images of C2C12 cells seeded at indicated densities in high-serum medium. Scale bar: 200 μm. (**b**) Differentiation efficiency of C2C12 cells at different seeding densities measured by the percentage of MyHC-positive cells. Data represent means ± SEM; $n = 3$ independent biological repeats. Statistical comparison was performed using the Mann–Whitney test and was non-significant

thus impact the results and reproducibility (Fig. 2). We recommend verifying the correct cell confluency by microscopy. Furthermore, the passage number also impacts the extent of differentiation [3]. We therefore strongly recommend utilizing the same passage numbers when performing biological repeats. In this study, we consistently use cell batches with passage number nine. Remember to retain the double-negative (i.e., GFP−/MyHC−) control sample under high serum conditions (*see* **Note 1**). Hereafter, continue with Sect. 3.5.

1. Cells should be at a confluency of approximately 90% at the time of the serum switch.
2. Aspirate the high-serum medium and add 2 mL of low-serum medium. For optionally transfected cells, do this 24 h after transfection.
3. The medium should be exchanged with fresh low-serum medium daily for 3 days to allow cells to differentiate into myotubes.

3.4 Cell Differentiation with Drug Treatment

This section is optional and outlines the steps to assess the impact of drug treatment on differentiation. While AMG 510 is used as an example here, this protocol can be applied to test any other drug or experimental compound, as illustrated by us before [3]. Do not add low-serum medium to the double-negative control sample. Hereafter, continue with Sect. 3.5.

1. Prepare a working solution of AMG 510 at a final concentration of 3 μM in prewarmed low-serum medium.
2. Prepare a DMSO-only (solvent) control in low-serum culture medium at the same concentration as AMG 510. Observe that the final DMSO concentration in the medium may not exceed 0.2%.
3. 1 day after transfection, replace the high-serum medium in each well with 2 mL of prepared working solution.
4. Replace the working solution daily for three consecutive days (*see* **Note 4**).

3.5 MyHC-Immunolabeling of Cells and Flow Cytometry

After optional transfections, optional drug treatments, and induction of differentiation, we describe in this section the immunolabeling with the Myosin 4 monoclonal antibody (MF20) conjugated to eFluor 660, i.e., the MyHC-antibody and flow cytometric measurements of target samples, MyHC+ control, and double-negative samples.

1. Aspirate the low-serum medium, add 1 mL 0.05% trypsin-EDTA, and incubate at 37°C until all cells fully detach from the flask surface (2–5 min).
2. Add 500 μL of complete medium to neutralize the 0.05% trypsin-EDTA and mix the cell suspension by pipetting up and down.
3. Transfer the cell suspension to a 1.5 mL centrifuge tube and centrifuge at 500 × *g* for 5 min at 23°C.
4. Remove the supernatant and add 500 μL of PBS.
5. Centrifuge at 500 × *g* for 5 min at 23°C.
6. Discard the supernatant and fix the cells with 500 μL 4% PFA at 23–25°C for 10 min. Mix thoroughly to prevent clumping.
7. After the incubation has ended, centrifuge again at 500 × *g* for 5 min.
8. Discard the supernatant and add 500 μL PBS to the pellet. Mix to resuspend the pellet. Samples can be stored at 4°C for up to 1 week at this stage.
9. Centrifuge the samples at 500 × *g* for 5 min at 23°C.
10. Aspirate supernatant and add 500 μL 0.5% Triton X-100 in PBS to the pellet for permeabilization, mix by vortexing, and incubate for 10 min at 23–25°C.
11. Centrifuge at 500 × *g* for 5 min.
12. Discard the supernatant and dissolve the pellet in 500 μL PBS + 0.05% Tween 20, mix by vortexing.
13. Centrifuge at 500 × *g* for 5 min and aspirate the supernatant.

14. Dilute Myosin 4 monoclonal antibody (MF20) conjugated to eFluor 660 (MyHC antibody), which detects MyHC protein, in PBS + 0.05% Tween 20 at a 1:100 dilution ratio.
15. Add the diluted MyHC antibody to the target samples, MyHC+ control, and double-negative control. For the GFP-variant positive control, skip this step and proceed with the same volume of PBS + 0.05% Tween 20 without the antibody.
16. Incubate the samples in the dark in the fridge at 4°C for 1 h.
17. Centrifuge at 500 × *g* for 5 min at 23°C.
18. Aspirate supernatant and resuspend the pellet in 500 μL PBS.
19. Acquire samples with the Guava easyCyte 6HT 2 L flow cytometer.

3.6 Channel Settings and Sample Acquisition with a Flow Cytometer

Here, the Guava easyCyte 6HT 2 L flow cytometer was used to acquire flow cytometry data. The instrument features 50 mW, 488 nm and 100 mW, 642 nm photodiode laser lines. The instrument configuration included the following filters: FSC and SSC (bandpass 488/16), Green-B (bandpass 525/30), and Red-R (bandpass 662/15). GFP-variant fluorescence was detected using 488 nm laser excitation and the Green-B filter. Other standard flow cytometers with these features can be employed.

1. Open the GuavaSoft software and launch the "Cleaning" module. Follow the onscreen instructions to prepare cleaning solutions by adding water and instrument cleaning fluid (ICF) in Eppendorf tubes as per on-screen instructions and place the tubes in the tube holder.
2. After cleaning the fluidic system, perform the easyCheck quality control procedure by preparing a 1:20 dilution of Guava easyCheck Beads (*see* **Note 5**).
3. The easyCheck quality control procedure is necessary for determining if the instrument optics, fluidics, and electronics are functioning as per the specified parameters.
4. When the instrument is ready, configure the acquisition settings to acquire 30,000 events per well, with an acquisition time of approximately 150 s in our case (*see* **Note 6**).
5. Load 300 μL of the control and the target samples into separate wells of a 96-well plate (*see* **Note 7**).
6. Begin acquisition with the double-negative control. Set up a dot plot with forward scatter channel (FSC-height/FSC-H) in linear scale on the x-axis and side scatter channel (SSC-height/SSC-H) in linear scale on the y-axis (Fig. 3a).
7. Debris and dead cells exhibit low SSC and FSC values and are found on the lower left corner of the plot (Fig. 3a).

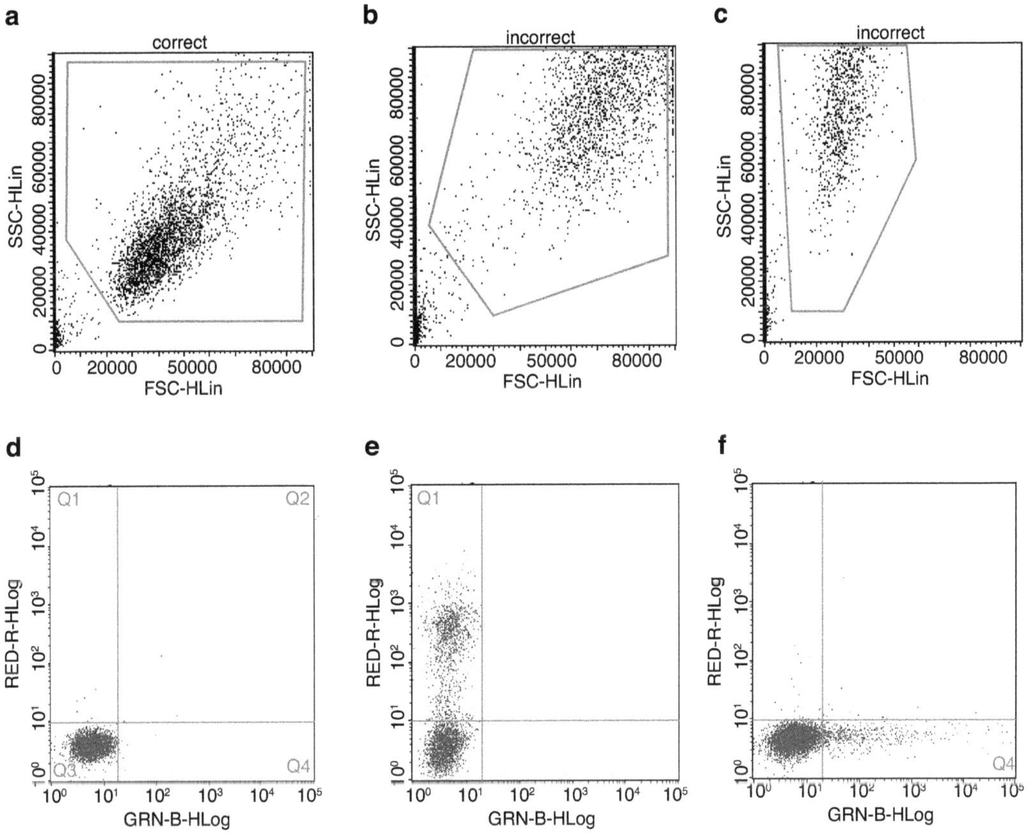

Fig. 3 Gating strategy for C2C12 cells in flow cytometry analysis. (**a**) With optimal FSC and SSC settings, the correct acquisition gate (indicated by the red polygon) places the main cell cluster into the center. (**b, c**) Incorrect FSC (**b**) or SSC (**c**) settings lead to a loss of larger, differentiated cells during analysis. (**d–f**) Adjust laser power and voltage gain to visualize double-negative control cells in Q3 (**d**), MyHC+ control cells in Q1 (**e**), and GFP-variant positive control cells in Q4 (**f**)

8. Adjust the FSC and SSC gain settings such that the major cell cluster appears distinctly in the center of the plot (Fig. 3a). This is an important step to ensure the acquisition of cells and must be stringently set up before an experiment. Suboptimal adjustment of FSC-H and SSC-H will lead to significant loss of the differentiated cell fraction during analysis (Fig. 3b, c) (*see* **Note 8**).
9. Draw a polygonal region around the cell cluster such that the debris is excluded (Fig. 3a), and gate the cells within this region onto a separate plot and visualize it as a dot plot using the Green-B versus the Red-R channels (Fig. 3d).
10. From the "Rregions" tab, "Quad stat marker" was selected to visualize the cells in four quadrants.

11. Adjust the laser power for the 488 nm and 642 nm lasers, along with the voltage gain for the Green-B and Red-R channels. The gain and laser power should be set so that the cluster of the double-negative control sample cells in Q3 appears just below 10^1 relative fluorescence units (RFU) in both channels. This ensures that all events in Q3 can be consistently identified as non-fluorescent cells (Fig. 3d). This step is essential to rule out false positivity, since even non-labeled and non-transfected cells can display background fluorescence.
12. After configuring the flow cytometer, verify that MyHC+ control cells appear along the vertical axis (Red-R channel) in Q1 (Fig. 3e) and GFP-variant positive control cells appear along the horizontal axis (Green-B channel) in Q4 (Fig. 3f).
13. After setting up the channels using the control samples, acquire all control and target samples.

3.7 Analysis of Cell Differentiation

After correct data acquisition, data are analyzed in four analysis quadrants generated using the "QuadStat Marker" window. For automated analysis of C2C12 cell differentiation data, we employ FlowFate, a free, interactive R-based tool developed by our group [4] (*see* **Note 9**).

1. Add a "New Stat" using the "Count" metric, which will allow quantification of cells in all four quadrants. Detailed procedures for visualizing cells in dot plots and quantifying them within quadrants or regions of interest can be found in all standard commercial flow cytometry analysis software.
2. Using the double-negative control as a reference, position the QuadStat Marker at 10^1 on both axes (Fig. 3d). This ensures that the cluster of cells below 10^1 in Q3 is correctly identified as non-fluorescent cells.
3. Cells are designated MyHC+ if the signal in the Red-R channel exceeds the 10^1 threshold in Q1 as established by the QuadStat Marker.
4. At this stage, the impact of an mEGFP-tagged gene of interest on C2C12 cell differentiation can be assessed. Cells with mEGFP fluorescence intensity between 10^1 and 10^2 RFU in the Green-B channel are categorized into the "GFP low" window (*see* **Note 10**). Within this window, cells are classified as MyHC+ if their signal in the Red-R channel exceeds a threshold of 10^1 RFU, which is determined using the QuadStat Marker (Fig. 4a, b).
5. For gene-dose-dependent analysis of mEGFP constructs, extra windows can be defined (*see* **Note 9**). Export the count of MyHC+ and MyHC− cells from the GFP low window to graphing and analysis software (e.g., Microsoft Excel)

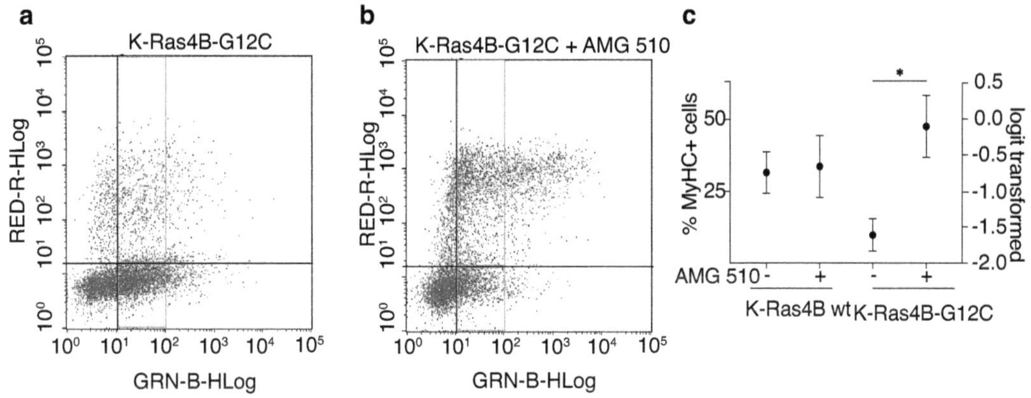

Fig. 4 Representative experiment illustrating the differentiation outcomes of cells transfected with K-Ras constructs and treated with K-RasG12C inhibitor AMG 510. (**a, b**) Plots showing MyHC-labeled cells expressing mEGFP-K-Ras4B-G12C treated with 0.2% DMSO in vehicle control (**a**) or mEGFP-K-Ras4B-G12C treated with 3 μM AMG 510 (**b**). (**c**) The percentage of MyHC+ cells analyzed in the 'GFP low' window (green box) reports on the differentiation. Plotted are means ± SEM of logit-transformed values to enable statistical evaluation using one-way ANOVA; *$p \leq 0.05$; $n = 5$ independent biological repeats

(*see* **Note 9**). Calculate the percentage of MyHC+ cells relative to the total number of GFP low population using the formula:

$$\%MyHC+ = \frac{MyHC+\ cells\ in\ GFP\ low\ window}{Total\ cells\ in\ GFP\ low\ window} \times 100$$

6. Transform the data by using the logit function to obtain continuous values for statistical comparison using one-way ANOVA. Employ the following formula: $\text{logit}(p) = \ln(p/1 - p)$, where p represents the fraction of MyHC+ cells (i.e., MyHC+ cells in GFP low window/total cells in GFP low window) (Fig. 4c).

4 Notes

1. Depending on the experimental objective, several types of samples have to be prepared: target samples (here, e.g., mEGFP-K-Ras4B-G12C and mEGFP-K-Ras4B wt), GFP-variant positive control samples. Additionally, we use non-transfected cells cultured under high-serum medium conditions as double-negative (i.e., GFP−/MyHC−) control sample, and as the MyHC+ control sample, we use non-transfected cells cultured in low-serum medium and processed in the same way as the target samples.
2. The amounts, as specified by the manufacturer, have been optimized for transfecting cells cultured in a six-well plate with

jetPRIME transfection reagent. Alternative transfection reagents may also be used with appropriate optimization. Some transfection reagents, such as Lipofectamine 2000, appear to impact differentiation.

3. Longer incubations with the transfection mix may cause cell toxicity.
4. The final concentration of DMSO should be kept below 0.2% in the medium to avoid unwanted effects on differentiation. If alternative solvents are used, the concentration that does not impact differentiation must be determined first.
5. The flow cytometer needs cleaning and unclogging of the microfluidics before usage, following the manufacturer's instructions. The easyCheck procedure calibrates the system and evaluates the sensitivity of optical components, including lasers and detectors. To ensure optimal system operation, fluorescence intensity and bead count must fall within the range specified by the manufacturer. When utilizing other flow cytometers, set up the equipment accordingly. There is no spectral overlap between GFP and eFluor 660; hence, fluorescence compensation was not necessary for our samples.
6. For highly concentrated cell samples, a shorter acquisition time may apply.
7. The Guava 6HT 2 L flow cytometer can acquire samples from either 1.5 mL Eppendorf tubes or 96-well plates. This procedure only covers sample acquisition using the instrument's 96-well plates, making it easier to acquire more samples and conserve sample volume.
8. Adjustment of FSC and SSC channels should be performed such that all cells are observed along a diagonal in the center of the plot. This ensures that larger cells, which exhibit high FSC and SSC values, are not excluded from the analysis. This is especially relevant for C2C12 cells, since the cells grow larger and undergo cell fusion upon differentiation. The resulting population is thus a mixture of smaller undifferentiated cells and larger differentiated cells. The FSC and SSC adjustment, as shown in Fig. 3a, ensures that the larger, differentiated cells that are observed toward the periphery of the plot are acquired. By contrast, if FSC and SSC gating is incorrectly set up, as shown in Fig. 3b, c, the acquisition of the larger, differentiated cells and, therefore, the MyHC+ cell fraction is reduced during analysis.
9. FlowFate processes flow cytometry data in flow cytometry standard (FCS) 3.0 format as individual or merged files. FlowFate integrates established R packages—flowCore, flowWorkspace, and ggcyto—into a customized workflow designed for differentiation data analysis. The software

quantifies differentiated cells across three GFP-expression windows, enabling gene-dose-dependent analysis of GFP-variant-tagged genes of interest. For example, the "GFP medium" window can be selected for mEGFP fluorescence intensity between 10^2 and 10^3 RFU in the Green-B channel, while the "GFP high" window comprises intensities between 10^3 and 10^4 RFU in the Green-B channel. These windows are used to identify MyHC+ and MyHC− cells. See our publication Parisi et al. for all details [4].

10. We recommend using cells within the "GFP low" window in our example, as it recapitulates results seen with endogenous protein expression levels [3]. This decision depends on the specific biological context and requires oversight by the user that cannot be covered here.

Acknowledgments

This work was supported by grants from the Luxembourg National Research Fund (FNR), AFR/23/17112420/Bil_ABANKWA_SPREDCanUL2 and INTER/FWO/23/18086068 molGluRAS2, to D.K.A.

References

1. Obinata M (2007) The immortalized cell lines with differentiation potentials: their establishment and possible application. Cancer Sci 98:275–283
2. Evano B, Tajbakhsh S (2018) Skeletal muscle stem cells in comfort and stress. NPJ Regen Med 3:24
3. Chippalkatti R, Parisi B, Kouzi F, Laurini C, Ben Fredj N, Abankwa DK (2024) RAS isoform specific activities are disrupted by disease associated mutations during cell differentiation. Eur J Cell Biol 103:151425
4. Parisi B, Sünnen M, Chippalkatti R, Abankwa DK (2023) A flow-cytometry-based pipeline for the rapid quantification of C2C12 cell differentiation. STAR Protoc 4:102637
5. Canon J, Rex K, Saiki AY, Mohr C, Cooke K, Bagal D et al (2019) The clinical KRAS(G12C) inhibitor AMG 510 drives anti-tumour immunity. Nature 575:217–223

Chapter 7

Light-Sheet Fluorescence Microscopy Imaging of Whole-Mount Immunostained Mouse Embryos and Limbs

Cristina Vicente-García, Alejandra Navarro-Martínez and Jaime J. Carvajal

Abstract

Animal development is the complex process that enables the formation of a pluricellular organism starting from a fertilized zygote. During this process, different tissues need to grow and differentiate in a coordinated manner in order to generate a fully functional organism. Oftentimes, technical or experimental difficulties cause imaging studies to focus specifically on the development of a specific tissue as if it were independent of the development of closely related structures. Further, classical analyses are performed on histological sections in two dimensions, which often only cover part of the tissue or organ under investigation. These analyses thus miss important information that can only be revealed in three-dimensional (3D) settings, which also allow the study of complete large structures. Here, we describe a method to perform whole-mount immunostaining and light-sheet fluorescence microscopy imaging of whole embryos and limbs up to the neonatal stage. While we focus on the co-development of skeletal muscle, tendons, and nerves, these techniques can be adapted to obtain high-resolution 3D images of large samples for the study of any other biological process.

Keywords Light-sheet fluorescence microscopy, Whole-mount immunostaining, Mouse, Embryonic development, Myogenesis, Limbs, Musculoskeletal system

1 Introduction

In the last decade, the scientific community has witnessed the birth of single-cell omics, molecular-based revolutionary technologies that enable the integration of several layers of molecular information (e.g., transcriptomic and epigenetic) in individual cells [1]. Most of these techniques, however, lack spatial information: where these cells are located within the tissue under investigation, and which interactions they can establish between one another. Spatial omics aim at solving this issue by preserving the spatial organization of cells, which is fundamental to fully understand the biological processes that are taking place [1]. One of the most remarkable current challenges is

Joe Quadrilatero (ed.), *Myogenesis and Skeletal Muscle Biology*, Methods in Molecular Biology, vol. 3057,
https://doi.org/10.1007/978-1-0716-5396-8_7,

to develop proper bioinformatic tools to integrate the very complex and vast amount of data these techniques generate [1].

While the deployment of these molecular approaches has greatly benefited the developmental biology field [2–5], they pose several important drawbacks. For example, their cost is still prohibitive for some laboratories, considerable bioinformatic expertise is required, and eventually, findings need to be functionally validated, usually using more classical methods.

In keeping with the maxim "seeing is believing," classical approaches in developmental biology before the molecular era include in situ hybridization, immunohistological staining, cell labeling, or lineage-tracing approaches. All these techniques require imaging, which has also experienced great advances in the last years, allowing the visualization of structures from the nano- to the mesoscale level with a high resolution.

In particular, light-sheet fluorescence microscopy (LSFM) is a very potent technique that allows the three-dimensional (3D) reconstruction of different types of samples, from cell cultures to organs and even whole organisms, while providing up to single-cell resolution and even allowing live imaging [6, 7]. All of this is particularly relevant in the field of developmental biology, where full understanding of how a tissue develops needs to integrate information regarding not only the tissue of interest but also cellular interactions with co-developing tissues, as well as dynamic processes such as cell migration or morphogenetic movements. Remarkable examples of the use of LSFM in developmental biology include early works focused in the in vivo early development of medaka, zebrafish, and fruitfly embryos [8, 9], with detailed analysis of development of the mouse face [10] being one of the latest contributions.

The basic principle of LSFM is simple: two illumination objectives located at either side of the sample illuminate a thin plane with a sheet of light (hence the name of the technique), while the detection objective, which is located perpendicular to the other two, collects the light from the focal plane. By acquiring images along the z-axis, a 3D image of the whole sample can be reconstructed. This 3D optical sectioning ability, shared with regular confocal microscopy, eliminates the need for actual physical sectioning when working with fixed samples, thus reducing experimental time. Furthermore, it prevents the destruction of the sample and allows the selection of more precise and even multiple sectioning angles to visualize different features after image reconstruction. Unlike regular confocal microscopy, LSFM allows sample imaging from different angles sequentially, thus greatly improving resolution after the integration of the different views with appropriate software. Further, low photobleaching and phototoxicity, coupled with high speed and resolution, as well as the possibility to image thick samples, are additional advantages of LSFM over regular confocal microscopy. Downsides, on the other hand,

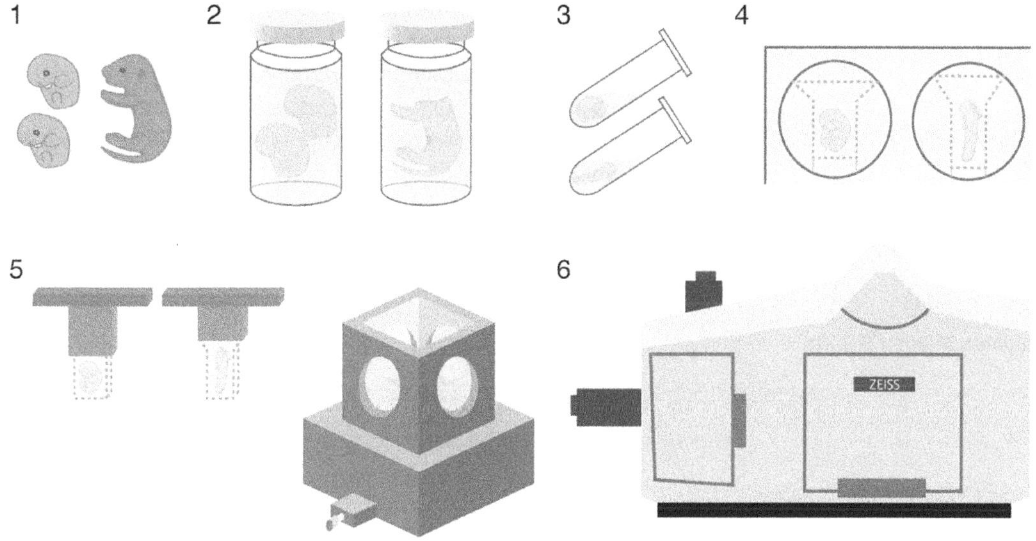

Fig. 1 Schematic representation of the whole-mount imaging of mouse samples procedure. Embryos or postnatal pups are collected (1), processed, bleached, fixed, and permeabilized (2). Limbs at the required stages are dissected, while early embryos are left whole. Samples are then immunostained (3) and embedded into agarose blocks (4). After clearing, blocks are glued to the light-sheet sample holder and introduced into the imaging chamber (5). Finally, the LSFM system is configured and imaging proceeds (6)

include the absolute need to make samples completely transparent with often toxic clearing agents and expertise in the use of dedicated software for image processing and 3D reconstruction.

Here, we describe a method to perform whole-mount immunostaining of mouse embryos, as well as fetal and neonatal limbs, coupled with LSFM, to study the development of mouse skeletal muscle (Fig. 1). Of note, the whole-mount immunostaining procedure has been adapted from the protocol previously reported by Kardon's laboratory [11]. While the 3D reconstruction of LSFM falls out of the scope of this manuscript, we suggest the reader to test different image analysis tools in order to find the one that better adapts to their samples in terms of software capacity to manage large file sizes, accuracy, ease of use, and functionalities. Examples of excellent image analysis software include commercial Arivis Pro (Zeiss) and Imaris (Oxford Instruments), as well as the free tools FluoRender [12] and Paintera [13].

2 Materials

2.1 Mouse Model

Tg(Scx-GFP) mice carry a transgenic construct that extends approximately 4 kb upstream and 5 kb downstream of the mouse *Scleraxis* gene, in which the GFP reporter gene was cloned into the first exon beginning with the initiator ATG of the endogenous gene and replacing most of this first exon [14]. This mouse line, kindly provided by Ronen Schweitzer, allows the visualization of all developing tendons and ligaments, since *Scx* is a key transcription factor in the development of these connective tissues. Perform housing and experimental protocols approved by the corresponding ethical committee in accordance with all relevant ethical, scientific, and legal guidelines.

2.2 Sample Collection and Fixation

1. 1× phosphate-buffered saline (PBS).
2. 4% paraformaldehyde (PFA) prepared in 1× PBS.
3. 92 × 16 mm transparent Petri dishes.
4. 7 mL glass screw-neck vials with screw caps (Fisher Scientific).
5. 20 mL glass screw-neck vials with screw caps (Sigma-Aldrich).
6. 2 mL microtubes.
7. Dissecting tools (all from Fine Science Tools): sharp fine scissors, Dumont forceps #4, Dumont forceps #5, Dumont forceps #5/45.
8. Silicone-coated dissection dish and pins.
9. Water bath.
10. Stereomicroscope (Nikon SMZ1500).
11. Shaker or roller.

2.3 Sample Bleaching, Fixation, and Permeabilization

1. 1× PBS.
2. 100% methanol.
3. Dent's fix solution: 1 part dimethyl sulfoxide (DMSO): 4 parts 100% methanol.
4. Dent's bleach solution: 1 part 30% (v/v) hydrogen peroxide: 2 parts Dent's fix solution.
5. 10 mL serological pipettes.
6. Glass pipettes.
7. Aluminum foil.
8. Shaker or roller.

2.4 Whole-Mount Immunofluorescence Staining

1. 1× PBS.
2. Blocking solution: 5% goat serum (Gibco), 75% 1× PBS, 20% DMSO.
3. Primary antibodies: mouse anti-skeletal myosin (1:400, MY-32, Sigma M 4276), rabbit anti-neurofilament L (1:100, NF C28E10, Cell Signaling #2837), chicken anti-GFP (1:200, Abcam ab13970), monoclonal ANTI-FLAG® M2 antibody (1:350, Sigma-Aldrich F1804).

4. Secondary antibodies (all Invitrogen and used at a 1:500 dilution): Alexa 405 goat anti-rabbit IgG (A31556), Alexa 555 goat anti-chicken IgY (A32932), Alexa 633 goat anti-mouse IgG1 (A21126).
5. 2 mL microtubes.
6. 10 mL serological pipettes.
7. Glass pipettes.
8. Surgical blades.
9. Dissecting tools, silicone-coated dissection dish and pins (see Sect. 2.2).
10. Shaker or roller.

2.5 Sample Embedding, Clearing, and Mounting

1. Agarose D1 LE (Low EEOO).
2. Ultrapure water.
3. BABB solution: 1 part benzyl alcohol (Thermo Scientific): 2 parts benzyl benzoate (Gelntham Life Sciences).
4. 100% methanol.
5. Petri dishes.
6. 20 mL glass screw-neck vials with screw caps (Sigma-Aldrich).
7. 40 μm nylon cell strainers.
8. 6-well flat-base cell culture plates.
9. Chattaway-type spatula.
10. Surgical blades.
11. Glass pipettes.
12. 25 mL glass beakers.
13. Aluminum foil.
14. Water bath.

2.6 LSFM Imaging

1. BABB for imaging.
2. Absolute ethanol.
3. Super Glue-3.
4. Glass Petri dish.
5. Glass Pasteur pipettes and pipettor.
6. ABBE analogue refractometer RST111 (Metria).
7. Zeiss Lightsheet 7 microscope with the ZEN 3.1 software.
8. Zeiss Lightsheet 7 microscope objectives: CE NEOFLUAR 5×/0.16 foc detection optics and 5×/0.1 foc illumination optics.
9. Zeiss Lightsheet 7 microscope accessories: T-shaped mounting stabs and large sample chamber ($n = 1.33$–1.58).

3 Methods

Unless specified, all steps are performed at room temperature. Dispose all biological and chemical waste according to safety regulations. Use appropriate protective equipment, including gloves and a lab coat, throughout the procedure. Take extra precaution when handling toxic reagents such as methanol or BABB.

3.1 Collection and Fixation of Embryos at the Desired Developmental Stage

1. Set up the appropriate mating between male and female mice to obtain embryos with the desired genotypes (*see* **Notes 1 and 2**).
2. On the next morning, check for the presence of the viscous vaginal plug in each female. If present, copulation has taken place, the female is probably pregnant, and that day is considered as 0.5 day postcoitum (dpc) (*see* **Note 3**).
3. Sacrifice the pregnant female by cervical dislocation at the desired stage of embryo development. Remove the bicornuate uterus and pin both ends of one of the horns to a silicone-coated dissection dish filled with cold PBS making sure it is well stretched (Fig. 2a).

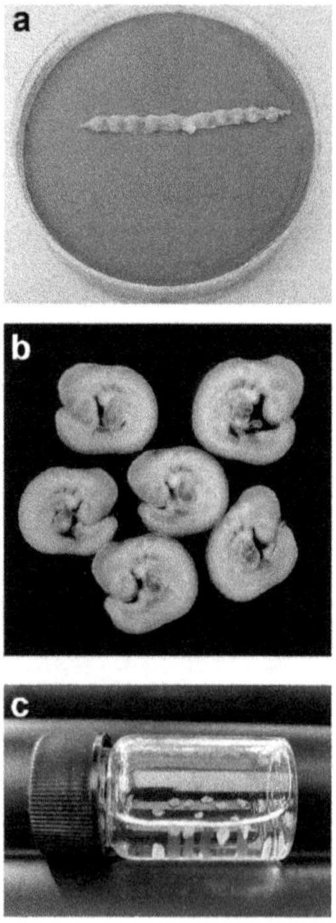

Fig. 2 Collection of 9.5 dpc mouse embryos. (**a**) After removing the uterine horns from the peritoneal cavity, pin them in a silicone-coated dissection dish with PBS, (**b**) dissect embryos, and (**c**) fix them in 4% PFA/PBS-containing glass vials

4. Carefully dissect the embryos under a stereomicroscope using micro scissors and fine forceps: cut the uterus exposing the individual embryos, pierce and tear apart the surrounding decidua, cut the umbilical cord, and remove the extraembryonic membranes (Fig. 2b).
5. Fix embryos in 7 mL glass vials filled with 4% PFA/PBS overnight, rocking or rolling at 4°C. If samples need to be genotyped, fix them individually in numbered glass vials (Fig. 2c). In parallel, carefully dissect their extraembryonic membranes to avoid cross-contamination and place them individually in 2 mL microtubes labeled with the same number as their corresponding embryo. Extract genomic DNA from these membranes and genotype as appropriate (*see* **Note 4**).
6. Repeat the procedure with the other uterine horn.

3.2 Collection and Fixation of Postnatal Samples

1. After setting appropriate matings (see Sect. 3.1, step 1), collect pups on the day of birth (postnatal day 0 or P0), typically 20–21 days later.
2. Sacrifice them carefully by decapitation without disturbing the shoulders and wash them briefly in water to remove blood (Fig. 3a).

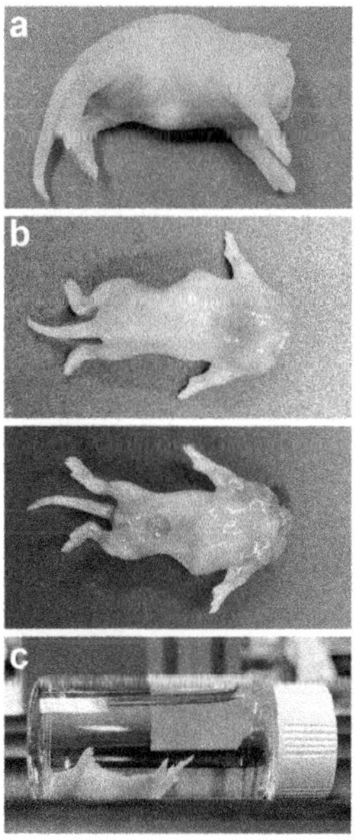

Fig. 3 Collection of newborn mouse pups. (**a**) Sacrifice P0 pups by decapitation, which is considered the most effective and humane means of euthanasia at this stage. (**b**) Skin and eviscerate them for successful fixation and immunostaining. (**c**) Fix samples individually in 4% PFA/PBS-containing glass vials

3. Remove the skin (Fig. 3b). To do so, first scald samples in hot water at 65°C for 30 s. Then, transfer them to a Petri dish filled with cold PBS under a stereomicroscope. Use micro scissors to make an incision along the back of each sample, starting at neck level. Then, slowly peel the skin away with fine forceps. Rotate the sample as necessary to remove the skin from the limbs. Proceed with caution at the most distal ends to prevent the undesired removal of digits (*see* **Notes 5 and 6**).
4. Eviscerate samples by making a small incision with micro scissors in the abdomen and using fine forceps. Further remove lungs and the heart from the chest cavity by first pulling out the diaphragm (Fig. 3b).
5. Place samples in 4% PFA/PBS-containing numbered 20 mL glass vials, rocking or rolling at 4°C overnight (Fig. 3c). If necessary, use the corresponding heads or the tip of the tail for genotyping.

3.3 Sample Bleaching, Fixation, and Permeabilization

1. Regardless of the type of sample (see Sect. 3.1 and 3.2), perform three quick washes with cold PBS to remove all residual PFA.
2. Remove the PBS and fill vials with Dent's bleach. Cover vials with aluminum foil to protect from light and incubate overnight in an upright position rocking at 4°C (see **Notes 7–9**).
3. Remove Dent's bleach with a glass pipette, fill vials with 100% methanol, close them, and rinse samples by inverting the vials five times. Repeat this washing step with methanol four more times.
4. After the last wash in methanol, fill the glass vials with Dent's fix and incubate samples rocking at 4°C (Fig. 4a). The minimum amount of time required for proper fixing and permeabilization of embryos, fetuses, and postnatal samples is 1 week, 2 weeks, and 1 month, respectively (*see* **Note 10**).

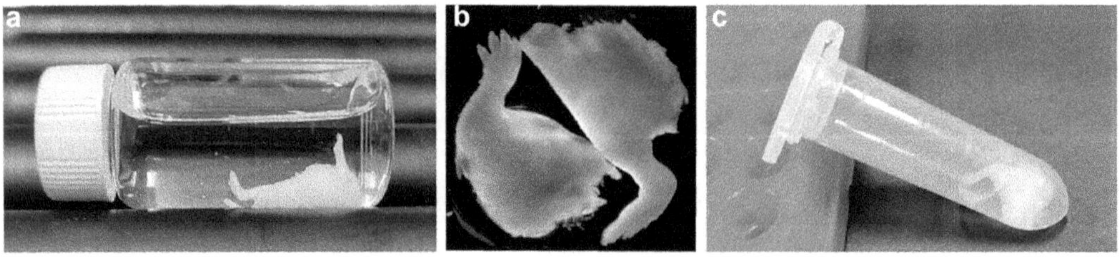

Fig. 4 Sample processing and whole-mount immunostaining. (**a**) Samples become opaque white after treatment with Dent's bleach. (**b**) Fetal and postnatal limbs are dissected after fixation and permeabilization with Dent's fix, but before immunostaining. (**c**) During the whole-mount immunostaining procedure, ensure samples are always completely covered in buffer

3.4 Whole-Mount Immunofluorescence Staining

1. Remove Dent's fix with a glass pipette and fill vials with PBS. Close the vials, rinse samples by inversion five times, and remove PBS with a plastic pipette.
2. Perform two additional quick washes, followed by two 1-h washes with PBS rocking or rolling in the same vials.
3. Dissect limbs from 14.5 dpc fetuses and postnatal samples in a silicone-coated dissection dish filled with cold PBS under a stereomicroscope. For this purpose, pin each sample with the dorsal side up and use a blade to cut through the midline in between the scapulas and the hip, thus cutting each sample in a sagittal plane. Using forceps and micro scissors, remove all tissue around fore- and hindlimbs (Fig. 4b). Embryos can be left whole.
4. Place up to two dissected limbs from fetuses and postnatal samples or one whole embryo in 2 mL microtubes and wash with PBS one additional hour, rocking or rolling.
5. Block samples for 1 h in 300 500 μL of blocking solution, gently rocking (Fig. 4c) (*see* **Note 11**).
6. Replace the blocking buffer with 300–500 μL of a primary antibody cocktail in blocking solution for three nights, gently rocking (*see* **Note 11**).
7. Perform three quick washes, followed by five 1-h washes in PBS.
8. Remove the PBS from the last wash and add 300–500 μL of a secondary antibody cocktail in blocking solution for three nights, gently rocking. Keep samples in the dark by wrapping tubes in aluminum foil or keeping them inside a dark opaque box on the rocking device (*see* **Notes 11 and 12**).
9. Perform three 5-min washes, followed by three 1-h washes in PBS. Protect from light.

3.5 Sample Embedding, Clearing, and Mounting

1. Prepare a 1% solution of agarose in water. Once the agarose is melted, keep the solution in a water bath at 60°C for 30 min to cool it down prior to filtering through a 0.40 μm cell strainer to remove potential floating particles (*see* **Notes 13 and 14**).
2. Take the agarose solution out of the water bath, pour 7 mL into one well of a six-well plate, and put the bottle back in the water bath. Cover the plate to prevent dust deposition. After 4 min, the agarose solution in the plate should have cooled down to 30°C (*see* **Note 15**).
3. Meanwhile, briefly wash one sample in a Petri dish filled with water to remove any trace of salts. Recover it gently with curved-tip forceps and remove excess water with Kimwipe tissue paper (*see* **Note 16**).

4. Place the sample gently in the middle of the agarose-containing well in the desired orientation. Ensure it is completely covered in agarose. Cover the plate to prevent dust deposition and protect from light (Fig. 5a) (*see* **Notes 17 and 18**).
5. Repeat steps 2 through 4 with as many samples as required (*see* **Note 19**).
6. Once the agarose is completely set, either preserve samples for future use (see step 7) or continue the procedure (see step 8) (*see* **Note 20**).
7. For sample preservation, pour some water on top of the wells to avoid desiccation of the agarose gels, cover the plate, protect it from light, and store at 4°C.
8. Trim an agarose block containing each specimen to be imaged using a sharp clean blade. Make the incisions up to the dish wall to ease the removal of the spare edges. Remove as much agarose as possible to reduce block weight (*see* **Note 17**). Ensure that the long axis of the sample will be aligned with the axis of rotation to minimize the light path through the samples. To be able to distinguish the top and bottom of each

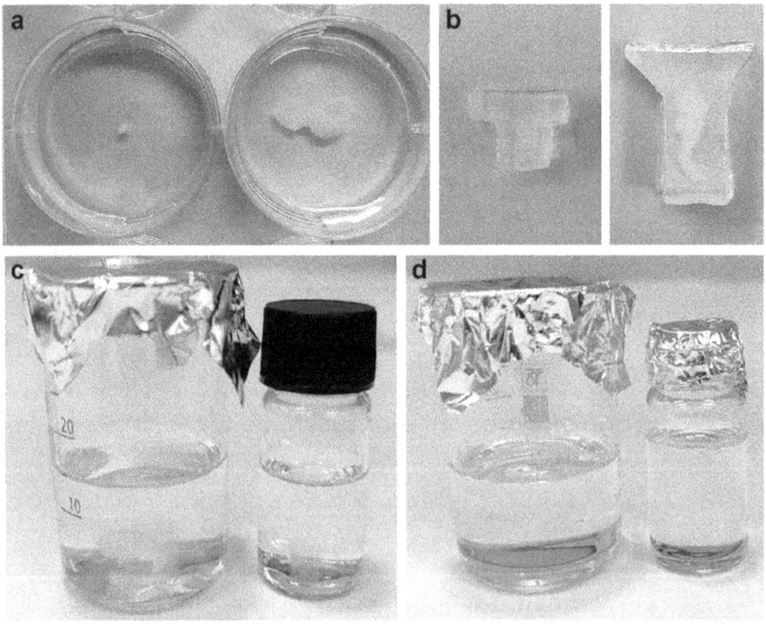

Fig. 5 Sample embedding and clearing. (**a**) Place immunostained samples in the center of a six-well plate well filled with 1% melted agarose. (**b**) Once set, trim an agarose block with the minimum possible amount of agarose in a rectangular fashion but with a distinctive shape at the top as a reference for sample position after clearing. Remove the spare edges, and (**c**) place each block in a glass beaker or vial filled with methanol. They should be covered with aluminum foil to protect from light (not shown). Overnight dehydration turns the blocks opaque. (**d**) Methanol is replaced with BABB, which makes samples transparent

sample after clearing, especially if blocks are rectangular, trim the top part with a distinctive shape (Fig. 5b).

9. Insert a Chattaway-type spatula in the incisions made with the blade and remove the spare edges carefully not to disturb the sample-containing block.
10. Carefully detach the blocks with the spatula and place them in 25 mL appropriately labeled glass beakers or vials (*see* **Note 21**).
11. Fill the beakers or vials with 100% methanol to dehydrate the samples, cover them with aluminum foil to protect from light, and keep them in the fume hood overnight.
12. On the next morning, remove methanol with a glass pipette, add fresh methanol, and incubate for eight additional hours (*see* **Note 22**).
13. Remove the methanol with a glass pipette and replace it with BABB, also using a clean glass pipette. Make some holes in the aluminum foil to allow residual methanol to evaporate and incubate in the fume hood overnight (*see* **Note 23**).
14. On the next day, thin samples should have become completely transparent and ready for imaging, while thicker ones may still be visible and would require additional incubation in BABB (Fig. 5d). Then, either store at 4°C for sample preservation or further clearing, or proceed with imaging (*see* **Note 24**).
15. Using a Chattaway-type spatula or flat wide tweezers, place an agarose block in a glass Petri dish. The top of the sample will be recognizable because of its distinctive shape.
16. Use a clean blade to make an incision perpendicular to the sample long axis in order to remove as much agarose as possible from the top and to expose a relatively dry surface.
17. Put a T-shaped mounting stub upside down in a clean glass Petri dish (Fig. 6a) (*see* **Note 25**).
18. Add glue on the mounting stub, place the top of block on it, and firmly press for around 1 min. Do it carefully not to deform the block. Then, wait for additional 4–10 min. Protect from light (*see* **Notes 26 and 27**).

3.6 Initialization of the LSFM System and Adjustment of the Refractive Index

1. Turn on the Zeiss Lightsheet 7 microscope and the computer.
2. Start the ZEN 3.1 software. Press the "Start system" button and wait until the system initializes and checks the status of all components. Specify the correct detection objective when required.
3. Measure the refractive index of BABB using a refractometer. To minimize optical aberrations, match the measured

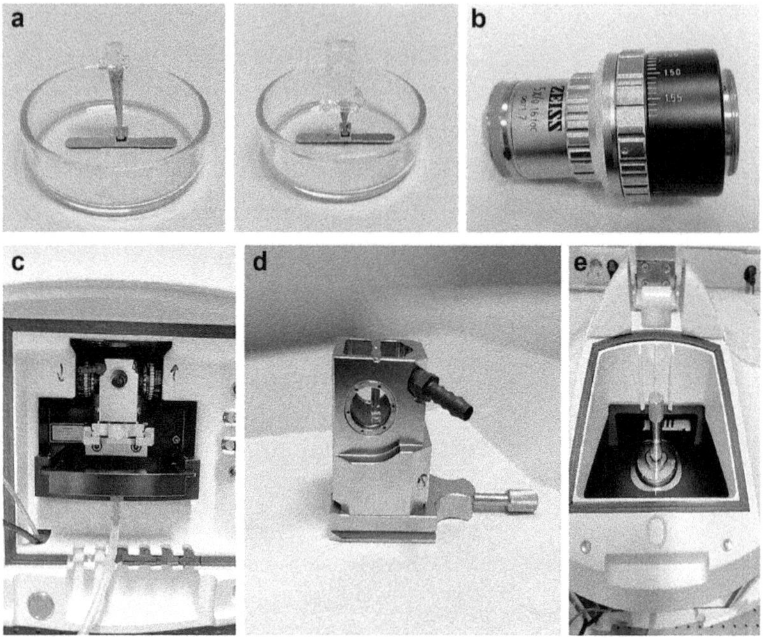

Fig. 6 Sample mounting for LSFM imaging. (**a**) Remove as much agarose from the top of the block as possible before gluing it to the light-sheet T-shaped mounting stub. (**b**) Meanwhile, set the correct refractive index in the 5× objectives with the adjustment collars (only the detection objective is shown) (**c**) and introduce the imaging chamber filled with BABB in the microscope. After adjusting the light sheet, (**d**) take out the chamber to introduce the mounting stub with the block glued to it, (**e**) insert the claw of the holding device in the microscope, and follow the software instructions for sample docking

refractive index with that of the 5× detection objective using the adjustment collar (Fig. 6b). Mount the objective into the microscope (*see* **Note 28**).

4. Adjust the refractive index of the pair of 5× illumination objectives and install them perpendicular to the detection objective (*see* **Note 29**).

3.7 Adjustment of the Light Sheet

1. Fill the large imaging chamber with enough BABB to cover the port windows and making sure there is no leakage.
2. Introduce the chamber in the microscope through the front system cavity door and fasten the screw finger-tight to lock it in place (Fig. 6c). Close the door.
3. There are four main tabs in the upper left corner of the software window: "Locate," "Acquisition," "Processing," and "Maintain." Click on the "Maintain" tab and make sure the right objectives are specified in the "Objectives" menu.

4. In the "Maintain" tab, expand the "Adjustment" dropdown menu and click on the "Adjust Light Sheet" button to focus the two illumination objectives. Select the laser line with the lowest wavelength that will be used (e.g., 405 nm), as well as the refractive index.
5. In the pop-up window, use the y-slider to overlay the light sheet with the green horizontal line that serves as a guide in the software. If no light sheet is visible, it will be necessary to click on the "Find light sheet" button for broad focusing before changing to the "Fine Adjustment" option.
6. Focus the light sheet with the z-slider.
7. Select the "Illumination Right" button. Open the front system cavity door and manually adjust the focus ring of the corresponding illumination objective so that the thinnest part of the light sheet (waist) is in the center of the target marked with the blue vertical line. Move the focus ring clockwise to shift the waist to the left side or vice versa (*see* **Note 30**).
8. Record the number for the z-focus offset.
9. Repeat steps 7 and 8 for the other illumination objective after selecting the "Illumination Left" button.
10. Press "Finish" to leave the "Adjust Light Sheet" setup.

3.8 Sample Docking

1. Carefully take out the imaging chamber from the microscope.
2. Once the agarose block containing the sample is completely glued to the holder, carefully take it, turn it into the upright position, and place it onto the grooves of the imaging chamber (Fig. 6d).
3. Place the chamber back into the microscope and lock it in place.
4. Insert the claw of the holding device through the small upper system cavity door (Fig. 6e).
5. Under the "Locate" tab in the software, click on the "Docking" button in the "Specimen Navigator" menu. Follow the instructions to automatically pick up the mounting stub with the claw. Then, manually lock the claw through the small door on top of the microscope as guided by the software.
6. Click on "Overview" to position the sample in the center of the imaging field in bright field with the joystick. Then, use the "Locate sample" option, which emits infrared light, to fine tune sample positioning.
7. Add an approximate specimen diameter in the "Specimen Navigator" menu under the "Acquisition" tab to prevent collisions with the chamber walls.

3.9 Configuration of Lasers and Channels

1. Under the "Acquisition" tab, in the "Laser" menu in the "Setup Manager" section, enable the laser lines that are going to be used, which, in the case of the protocol detailed here, are in the blue (405 nm), red (561 nm), and far-red (638 nm) spectra (*see* **Note 31**).
2. In the "Channels" menu in the "Acquisition Parameter" section, configure "Track 1" by clicking on laser 405 nm and selecting the desired pseudocolor to be shown in the image. Introduce the values of the z-focus offset for the right and left illumination objectives obtained in Sect. 3.7, step 8. Also, set a preliminary exposure time of 20 ms and 10% of laser power.
3. Add two additional tracks and configure them for lasers 561 nm and 638 nm as in the previous step.
4. In the "Light Path" menu in the "Setup Manager" section, select the appropriate filters for each track, specifically: track 1 (laser 405 nm), SBS LP 490 dichroic filter BP 420–470; track 2 (laser 561 nm), SBS LP 560 dichroic filter BP 575–615; track 3 (laser 638 nm), SBS LP 640 dichroic filter LP 660. Make sure to enable the "Camera" in all tracks (*see* **Note 32**).
5. Focus the light sheet for each track. In the case of track 1, under the "Acquisition Mode" menu in the "Acquisition Parameter" section, select "Left" and "Single side." Then, click on the "Continuous" button to start visualizing the sample. With the joystick, select a plane that is close to the detection objective and that can be used as a reference to align both left and right objectives (*see* **Note 33**).
6. Stop image acquisition and switch to the "Right" illumination objective and check the image. It should be the same as for the "Left" objective. Otherwise, modify the offset values in the "Channels" menu. Repeat this step switching between the "Right" and "Left" objectives and modifying the corresponding offset values as required.
7. Select the "Dual side" option in the "Acquisition Mode" menu and click on the "Continuous" button. Alternating images detected after illumination with the right and left objectives will appear continuously. Make sure that the images are as similar as possible in terms of morphological features. Otherwise, keep on modifying the offset values in the "Channels" menu (*see* **Note 34**).
8. Select the "Online Dual Side Fusion" option and then "Mean" as the value for fusing the outcome of both sides. Activate the "Pivot Scan" option to minimize shadows and also the "Stationary" option, which prevents the movement of the scanner-lens in the y-axis to improve the field of view along this axis at the expense of a longer acquisition time (*see* **Note 35**).

9. Repeat steps 5–8 for each track.
10. Visualize the image under the "Continuous" mode and adjust laser exposure time and power for all laser lines until obtaining appropriate images without saturated pixels and with an optimal dynamic range according to the histogram.

3.10 Configuration of the Z-Stack and Multiview

1. Activate the "Z-stack" option in the "Experiment Manager" section. A new dropdown menu called "Z-Stack" will appear in the "Multidimensional Acquisition" section. Set the first and last plane to be imaged, as well as "Optimal" to automatically adjust the most appropriate slice thickness. This configuration will be used in all tracks.
2. Activate the "Multiview" option in the "Experiment Manager" section. In the corresponding dropdown menu, add as many angles as required to image the sample from multiple points of view. Use the rotation slider of the "Specimen Navigator" menu or directly write the desired angle. Select the "Center of Image" option for sample rotation (*see* **Notes 36 and 37**).
3. Activate "Slow Drive" in the "Specimen Navigator" menu for a slow rotation of the sample.
4. Define a new Z-stack for each new angle. Center the sample in the imaging field if necessary, moving it only in the horizontal plane. Do not change the vertical axis as it will disturb the integration of the information gathered from all angles to reconstruct the final image.
5. Set 3 s as the delay time to start imaging after the system rotates the sample between angles (*see* **Note 38**).

3.11 LSFM Imaging, Shut Down, and Clean-Up Procedures

1. Select "Start Experiment" and save the results in the appropriate directory using standardized nomenclature if possible (*see* **Note 39**).
2. Under the "Locate" tab in the software, click on the "Undocking" button in the "Specimen Navigator" menu. Follow the instructions to automatically release the mounting stub from the claw. Then, manually release the lock of the claw through the small door on top of the microscope.
3. If more samples need to be imaged, take out the chamber and replace the agarose block. Otherwise, continue with the following steps to shut down the system.
4. Exit the software, shut down the computer, and switch off the microscope.
5. Remove the sample from the chamber. With a glass Pasteur pipette, remove the BABB. As it is unmixable with water, eliminate all BABB residues using organic solvents such as ethanol. Further clean the chamber with water. Dry it and put it back inside the microscope for the next user.

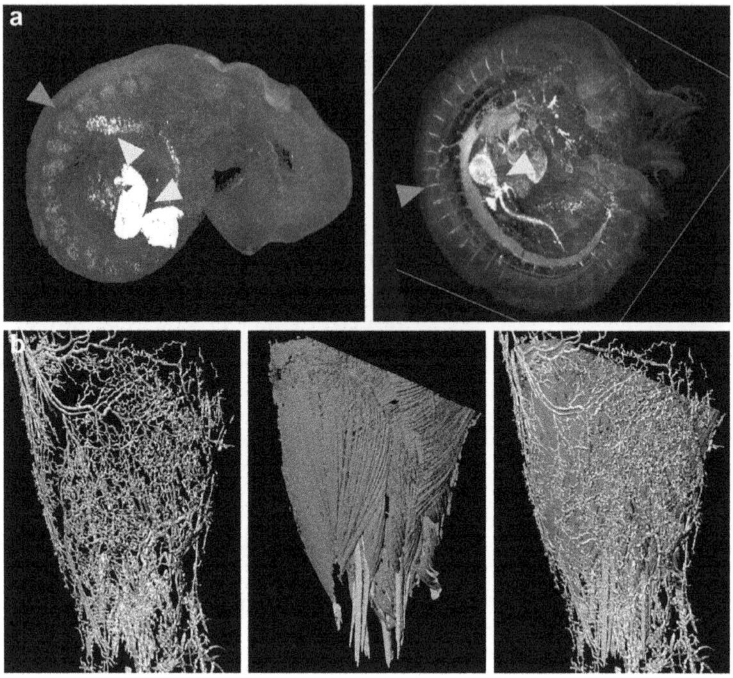

Fig. 7 Image analysis. (**a**) Examples of mouse embryos showing the expression of a skeletal muscle-specific marker throughout the somites at 9.5 dpc (left, red arrowhead) and myotomes at 10.5 dpc (right, red arrowhead). Autofluorescence is marked with yellow arrowheads. Images were processed with the ZEN 3.4 blue edition. (**b**) Three-dimensional reconstruction of skeletal muscles (magenta), tendons (green), and nerves (white) of a neonatal hindlimb using Imaris

6. Use a blade to remove the sample from the mounting stub. Clean the mounting stub as in the previous step and discard the sample.
7. Images can later be processed and analyzed with different tools that allow the segmentation and annotation of morphological features (Fig. 7).

4 Notes

1. Mate one male mice with one or even two females to increase the number of embryos or pups.
2. Mating efficiency can be increased by exposing the female to male pheromones by introducing bedding from the male's cage into the females' cage 3 days prior to mating. This results in the synchronization of the females' estrous cycles in such a way that most of them will be in the estrus phase and receptive to mating.
3. The vaginal plug consists of components of the male ejaculate and persists in the female's vagina for a limited amount of

time before falling out or dissolving. To prevent missing the presence of the plug, make sure to check females as early in the morning as possible.

4. Sometimes, embryos from the same mother are not at the same exact developmental time point. Stage young embryos by counting the number of somites, whereas older ones can be staged by the shape of the limb buds or limbs [15]. Make sure to use only specimens at the same developmental stage to avoid misleading conclusions.
5. Skin will act as a barrier to antibody penetration and needs to be removed for successful immunostaining.
6. This protocol has been optimized for postnatal limbs and fetuses up to 14.5 dpc. Older fetuses, especially after 16.5 dpc, would require skin removal as well, but scalding may not be necessary if skin is not fully developed yet.
7. Bleaching is part of the clearing process necessary to remove natural pigments in order to obtain optically clear samples for efficient imaging. Dent's bleach makes tissues opaque white. Expect the appearance of bubbles, especially from large samples, so make sure not to fully close the vials until all gases are gone.
8. Hydrogen peroxide, a component of Dent's bleach, is light sensitive, so protect your samples from light.
9. Always work under a fume hood when using methanol-containing solutions because it is volatile and toxic by inhalation. Also, methanol can degrade some plastics, so always use glass pipettes to handle methanol unless your plastic pipettes are methanol-resistant.
10. The longer samples are in Dent's fix, the better antibody penetration is obtained.
11. Adjust the blocking buffer volume and place the microtubes in an acute angle to ensure samples are completely covered in buffer.
12. Even after extensive bleaching and clearing, we observe persistent autofluorescence upon excitation at 488 nm, so we recommend avoiding secondary antibodies coupled with GFP or similar.
13. Low melting point agarose is preferable because it solidifies at a lower temperature than regular agarose, thus reducing the risk of damaging samples during embedding. However, regular agarose can be used as well.
14. Alternatively, the agarose solution can be filtered with a regular 0.45 μm filter and a syringe, but temperature is critical: if too high, the filter can melt, whereas if too low, agarose can solidify quickly, clotting the filter and making it separate abruptly from the syringe.

15. It is very important that agarose is cool enough not to damage samples, but still melted to allow sample embedding.
16. This step should be done as quickly as possible to prevent fluorescent dye degradation.
17. Embedding samples in agarose is required to produce a solid block that can be attached to the light-sheet sample holder. The bigger the block, the heavier it will be. Heavy blocks are hard to keep glued to the holder for a long period of time, so use the minimum amount of agarose on top and beneath the sample to reduce weight.
18. Sketch sample orientation in the laboratory notebook: after clearing, sample will be transparent and thus not visible any longer.
19. The remaining filtered 1% agarose solution can be reused. Simply melt it again for the next experiment.
20. Agarose is completely set when it acquires a translucent tone.
21. Pour a little bit of water in the wells to detach the blocks if necessary.
22. Afterward, samples should be completely dehydrated, acquiring an opaquer tone (Fig. 5c). Otherwise, change methanol again and incubate for a longer period of time.
23. BABB is highly toxic and corrosive to some plastics. It is recommended to place beakers on a glass tray to be able to constraint potential spillages or drops in order to safely clean them.
24. Clear postnatal limbs for 1 week before imaging.
25. Mounting stubs come in different sizes. Use the stub that would position the sample in the middle of the imaging chamber, allowing the visualization of the complete sample. Large samples such as postnatal limbs would require the smallest stub, whereas small samples such as young embryos would require a longer one.
26. Remember the mounting stub is upside down, so the top of the sample will be momentarily also upside down while gluing.
27. Avoid adding to much glue because it will run down the stub. Clean excess glue if possible because it would block the entry of the light beam in the specimen while imaging.
28. The refractive index of BABB would be around 1.56.
29. This is a preliminary adjustment as further adjustments will need to be made in the following section.
30. Since the laser will be off when the system cavity door is open, it will be necessary to close the door to check the position of the waist. Repeat this step as many times as needed.
31. While several laser lines could be used at the same time in the same track, in our hands, it works best to use one laser line

per track to enhance resolution, given that the z-focus offset is optimal only for one of the lines, at the cost of potential displacements in the z direction between channels.

32. In all cases, the primary laser blocking filter (LBF) is the same (LBF 405/488/561/640). The secondary beam splitter (SBS) and bandpass (BP) or longpass (LP) dichroic filters need to be selected according to the laser lines that are going to be used.
33. It might be necessary to adjust the zoom for large or small samples. In the case of very large samples, such as postnatal limbs, the user has to decide whether to take images of selected areas (e.g., just the paw) or configure the "Tiling" tool in the "Experiment Manager" section.
34. Signal intensities may vary because laser beams come from opposite sites.
35. The "Maximum" value for image fusion, or not using the "Stationary" option, could work best for some samples, so users should try all configurations.
36. The more the number of views, the longer the imaging process will take and the larger the file size will be. It is recommended to add one more angle, which should be perpendicular to the view already established.
37. Tiling is not possible when more than one angle is imaged.
38. If more than one angle is set and the microscope has to rotate the sample, the lower part of large blocks, which is far away from the attachment point, inevitably moves by inertia. Setting a delay time in imaging after rotation helps to overcome this issue.
39. Standardized nomenclature aids in the process of image storage and processing, especially when dealing with large batches of images. An example of standardized nomenclature would be: date_genotype_condition_target-channel_XX.czi. A batch of images obtained following the protocol described in this manuscript could then be: 20250401_WT_P0_NF-405_Scx-555_My32-633_01.czi, 20250401_WT_P0_NF-405_Scx-555_My32-633_02.czi, etc.

Acknowledgments

The authors thank Ronen Schweitzer for providing Tg(Scx-GFP) mice, as well as the animal and Advanced Light Microscopy and Image Analysis facilities at CABD, especially to Juan Carlos Millán Renquel, for their technical support. This work was supported by grants from the Spanish Ministerio de Ciencia, Innovación y

Universidades (PID2023-148397NB-I00), as well as from Junta de Andalucía (ProyExcel_00153, PAIDI 2021 call, and grant to the CABD Maria de Maeztu Excellence Unit).

References

1. Vandereyken K, Sifrim A, Thienpont B, Voet T (2023) Methods and applications for single-cell and spatial multi-omics. Nat Rev Genet 24:494–515. https://doi.org/10.1038/s41576-023-00580-2
2. Cao J, Spielmann M, Qiu X et al. (2019) The single-cell transcriptional landscape of mammalian organogenesis. Nature 566:496–502. https://doi.org/10.1038/s41586-019-0969-x
3. Regev A, Teichmann SA, Lander ES et al (2017) The human cell atlas. eLife 6. https://doi.org/10.7554/eLife.27041
4. Lohoff T, Ghazanfar S, Missarova A et al. (2022) Integration of spatial and single-cell transcriptomic data elucidates mouse organogenesis. Nat Biotechnol 40:74–85. https://doi.org/10.1038/s41587-021-01006-2
5. Argelaguet R, Clark SJ, Mohammed H et al. (2019) Multi-omics profiling of mouse gastrulation at single-cell resolution. Nature 576:487–491. https://doi.org/10.1038/s41586-019-1825-8
6. Weber M, Huisken J (2011) Light sheet microscopy for real-time developmental biology. Curr Opin Genet Dev 21:566–572. https://doi.org/10.1016/j.gde.2011.09.009
7. Keller PJ, Dodt H-U (2012) Light sheet microscopy of living or cleared specimens. Curr Opin Neurobiol 22:138–143. https://doi.org/10.1016/j.conb.2011.08.003
8. Keller PJ, Schmidt AD, Wittbrodt J, Stelzer EHK (2008) Reconstruction of zebrafish early embryonic development by scanned light sheet microscopy. Science 322:1065–1069. https://doi.org/10.1126/science.1162493
9. Huisken J, Swoger J, Del Bene F et al. (2004) Optical sectioning deep inside live embryos by selective plane illumination microscopy. Science 305:1007–1009. https://doi.org/10.1126/science.1100035
10. Lo Vercio LD, Green RM, Dauter A et al (2025) Quantifying the relationship between cell proliferation and morphology during development of the face. Development 152. https://doi.org/10.1242/dev.204511
11. Colasanto MP, Eyal S, Mohassel P et al. (2016) Development of a subset of forelimb muscles and their attachment sites requires the ulnar-mammary syndrome gene Tbx3. Dis Model Mech 9:1257–1269. https://doi.org/10.1242/dmm.025874
12. Wan Y, Otsuna H, Holman HA et al (2017) FluoRender: joint freehand segmentation and visualization for many-channel fluorescence data analysis. BMC Bioinformatics 18:280. https://doi.org/10.1186/s12859-017-1694-9
13. Paintera. https://github.com/saalfeldlab/paintera
14. Pryce BA, Brent AE, Murchison ND et al. (2007) Generation of transgenic tendon reporters, ScxGFP and ScxAP, using regulatory elements of the scleraxis gene. Dev Dyn 236:1677–1682. https://doi.org/10.1002/dvdy.21179
15. Kaufman, MH, 1992. The atlas of mouse development. Academic Press, London

Chapter 8

Ultrastructural Analysis of Enzymatically Dissociated Skeletal Muscle Fibers: Microwave-Assisted Heavy Metal Staining

Sergey Mursalimov, Ilan Zemski and Ori Avinoam

Abstract

In this chapter, we describe a protocol for the preparation of enzymatically dissociated murine skeletal muscle fibers for ultrastructural analysis. Enzymatic dissociation is a widely used method for the isolation of individual muscle fibers. At the same time, this treatment can induce significant physiological alterations, affecting sample characteristics and influencing the outcomes of bioimaging protocols. We found that the conventional transmission electron microscopy (TEM) tissue-processing protocol, performed at room temperature with resin embedding and heavy-metal staining, is unsuitable for samples of this type. Osmium tetroxide and uranyl acetate exhibited a drastically reduced penetration ability in dissociated fibers compared to skeletal muscle fragments or cultured myotubes, leading to a critical reduction in contrast in the resulting electron microscopy images. We used microwave-assisted tissue processing to overcome this obstacle. Besides ultrastructural analysis, the described protocol could be applied to other bioimaging approaches (such as immunostaining and related methods) that require overcoming penetration-related challenges when working with dissociated fibers.

Keywords Collagenase dissociation, Skeletal muscle fibers, Transmission electron microscopy, Resin embedding, Microwave processing

1 Introduction

Enzymatically dissociated muscle fibers are an important model system for investigating muscle function, development, and regeneration under defined experimental conditions. They are typically obtained by treating skeletal muscle fragments with collagenase, an enzyme that breaks down the extracellular matrix, allowing individual fibers to be isolated [1–4]. This approach preserves many functional aspects of mature skeletal muscle cells while providing a simplified and accessible system for a wide range of experimental manipulations, including transgene introduction, calcium measurements, and various pharmacological or imaging assays. However, collagenase treatment also

Joe Quadrilatero (ed.), *Myogenesis and Skeletal Muscle Biology*, Methods in Molecular Biology, vol. 3057,
https://doi.org/10.1007/978-1-0716-5396-8_8,

induces physiological and structural alterations in the fibers [2, 3]. These changes appear to be the main source of the challenges encountered when processing these fibers for electron microscopy.

In our ultrastructural study of murine dissociated fibers using conventional TEM [5, 6], we encountered challenges that did not arise in experiments with other muscle-related samples. Specifically, a critical reduction in contrast was noted in samples containing dissociated fibers (Fig. 1A). This could be explained by a striking loss of the ability of osmium tetroxide and uranyl acetate to penetrate dissociated fibers. In contrast, skeletal muscle fragments directly dissected from tissue, as well as cultured myotubes obtained from the same mice and treated with the same chemicals under identical conditions, consistently show sufficient contrast throughout the samples (Fig. 1B, C).

Even more significant is the observation that only healthy dissociated fibers, with well-preserved sarcomere structures, mitochondria, and membranes, exhibit problems during TEM processing. Typically, around 25% of dissociated fibers fail to withstand the preparation procedure and undergo pathological changes, losing their normal structure and turning into so-called hypercontracted fibers [7]. In our experiments, these hypercontracted fibers displayed excellent contrast after conventional TEM processing, even when present in

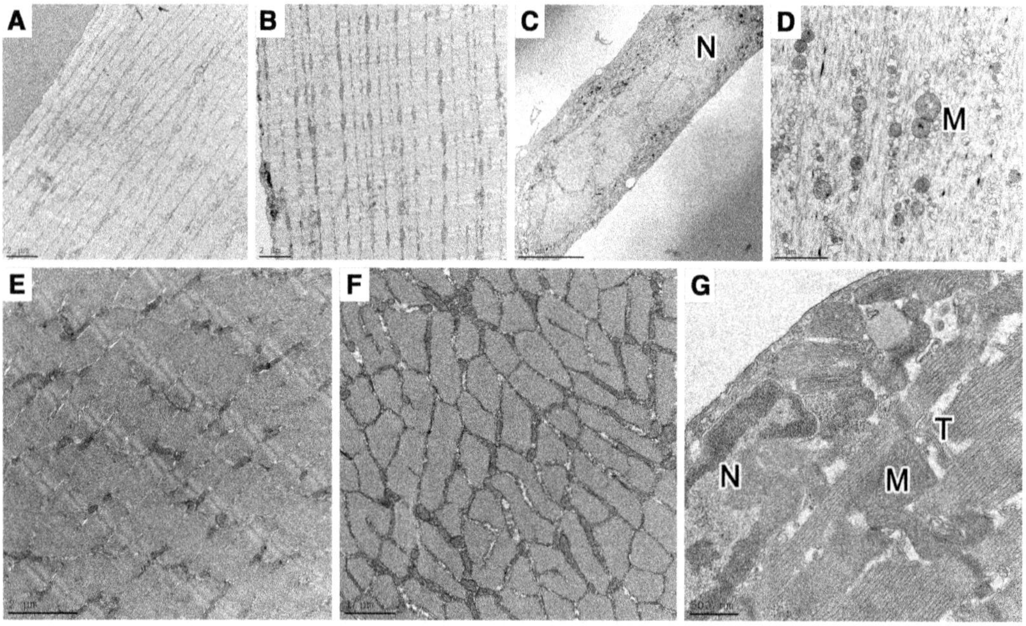

Fig. 1 Ultrastructural images of muscle-related samples prepared using conventional (**A–D**) and microwave-assisted (**E–G**) tissue-processing protocols for TEM. Healthy dissociated fiber showing poor contrast in (**A**). Skeletal muscle fragment in (**B**), myotube in (**C**), and hypercontracted fiber in **D**, all displaying good contrast. (**E and F**) Healthy dissociated fiber with good contrast, in different orientations in (**E**) and (**F**), and in higher magnification in (**G**). M, mitochondria; N, nucleus; T, T-tubules

the same tubes in which healthy fibers show no contrast. Unfortunately, along with excellent contrast, the hypercontracted fibers also exhibit sarcomere disruption, mitochondrial swelling, and breakdown of the sarcoplasmic reticulum membranes, making them unsuitable for analysis (Fig. 1D). Thus, only healthy, structurally intact fibers demonstrate poor penetration of heavy metals, while hypercontracted fibers stain effectively and can serve as an internal control to confirm that the contrast issue is not related to reagent quality or protocol parameters.

The obtained results allow us to conclude that neither muscle-specific features nor in vitro conditions are responsible for the observed changes in dissociated fibers. Instead, collagenase treatment appears to be the primary factor responsible for the loss of cellular permeability, which is essential for contrast agent penetration. It is known that cells can lose their physiological phenotype when the extracellular matrix is altered or disrupted, resulting in changes in gene expression patterns and organelle structure [3]. However, we found little information on how such alterations could block contrast agent penetration, especially of one as highly chemically reactive as osmium tetroxide.

To overcome these obstacles, we developed a protocol based on microwave treatment. The use of microwaves for tissue processing in microscopic analysis was first proposed almost 60 years ago and has since been incorporated into various protocols, including fixation, paraffin and resin embedding, immunolabeling, in situ hybridization, and other applications [8–10]. Microwaves significantly accelerate all processes within treated tissues by inducing rapid oscillation of water molecules and protein side chains, which leads to increased molecular movement and collisions throughout the sample. Primarily, this leads to a drastic reduction in processing time, but it also enhances the diffusion of chemicals into the tissue [9, 10]. However, despite their proven benefits, microwave-assisted techniques have been adopted much more slowly in electron microscopy and have not been widely incorporated into conventional EM protocols.

We found that microwave treatment significantly increases the uptake of osmium tetroxide and uranyl acetate by dissociated fibers, resulting in a sufficient level of contrast in the resulting ultrastructural images. At least half of the healthy fibers are suitable for analysis after the applied protocol (Fig. 1E–G). The fact that microwave treatment facilitates heavy metal penetration into dissociated fibers suggests that enzymatic dissociation directly impairs permeability, potentially by modifying cell surface characteristics or altering sample density. However, the exact mechanisms remain unknown.

Thus, the described TEM sample preparation protocol, which includes microwave processing, is suitable for ultrastructural analysis of enzymatically dissociated skeletal muscle fibers. With some modifications, this protocol could be adapted for use in other bioimaging tasks that require overcoming penetration-related challenges when working with dissociated fibers. It should also be noted that our results highlight the limitations of using enzymatically dissociated fibers as a model. While studies on isolated single muscle fibers provide valuable insights, researchers should be aware that they do not fully represent intact skeletal muscle fibers in vivo due to the physiological alterations induced by enzymatic dissociation.

2 Materials

2.1 Fixation

1. Prepare enzymatically dissociated murine (*Mus musculus*) muscle fibers according to one of the common protocols [1–4].
2. 1× PBS: mix 5 mL of 10× phosphate-buffered saline (PBS), pH 7.4, and 45 mL Milli-Q water.
3. 2× aldehyde fixative solution in media (5% glutaraldehyde and 4% paraformaldehyde): mix 1 mL of 25% glutaraldehyde solution, 1.25 mL of 16% paraformaldehyde solution, and 2.75 mL of media. Keep the solution on ice until use.
4. 1× aldehyde fixative solution in 1× PBS (2.5% glutaraldehyde and 2% paraformaldehyde): mix 1 mL of 25% glutaraldehyde solution, 1.25 mL of 16% paraformaldehyde solution, 1 mL of 10× PBS, pH 7.4, and 6.75 mL of Milli-Q water. Keep the solution on ice until use.
5. Microcentrifuge.
6. Standard laboratory plasticware (tubes, tips, pipettes, etc.).

2.2 Post-Fixation and Heavy Metal Staining

1. 3% potassium ferrocyanide in 2× PBS: mix 5 mL of 10× PBS, pH 7.4, and 20 mL Milli-Q water to prepare 2× PBS. Dissolve 0.3 g of potassium ferrocyanide in 10 mL of 2× PBS (*see* **Note 1**).
2. 2% reduced osmium tetroxide solution: mix equal volumes of 3% potassium ferrocyanide in 2× PBS and 4% osmium tetroxide aqueous solution.
3. 2% uranyl acetate aqueous solution.
4. PELCO BioWave® Pro+ Microwave Processing System.

2.3 Dehydration and Resin Embedding

1. 30%, 50%, 70%, 90%, and 96% and anhydrous ethanol.
2. Anhydrous acetone.
3. Agar 100 epoxy resin: prepare the resin mixture for medium or hard blocks according to the manufacturer's instructions.

3 Methods

3.1 Fixation

1. Transfer the well plates with dissociated fibers from the thermostat to the chemical hood and immediately add an equal volume of ice-cold 2× aldehyde fixative solution in media directly to the wells. Ensure that there is sufficient space in each well to accommodate the added volume (*see* **Note 2**).
2. Fix on ice for 2 h.
3. Carefully remove the solution from the wells and replace it with ice-cold 1× aldehyde fixative solution in 1× PBS.
4. Transfer the plates to 4°C and incubate overnight.
5. Transfer the fixative with fibers into 1.5 mL tubes using 1 mL pipette tips with cut tips.
6. Centrifuge the tubes at 4000 rpm for 5 min.
7. Carefully remove the supernatant.
8. Add 1× PBS to the tubes and incubate for 10 min at room temperature.
9. Centrifuge the tubes at 4000 rpm for 5 min and remove the supernatant.
10. Repeat the washing step two more times.

3.2 Post-Fixation and Heavy Metal Staining

1. After the final wash, remove the supernatant and add 1 mL of 2% reduced osmium tetroxide solution (*see* **Note 3**).
2. Do not close the tubes! Under the chemical hood, transfer them to a vacuum chamber. Close the chamber, place it in the microwave processor, and connect it to the vacuum port using a rubber tube inside the instrument.
3. Start the preinstalled "Osmium fixation" program, which consists of the following settings: 2 min of vacuum with microwaving (power: 100 watts), followed by 2 min of vacuum without microwaving. This cycle is repeated four times.
4. After completion, disconnect the vacuum chamber from the microwave and return it to the chemical hood.
5. Close the tubes and incubate them at room temperature for 30 min (*see* **Note 4**).
6. Without changing the solution in the tubes, repeat steps 2–4.
7. Transfer tubes under the chemical hood and carefully remove the osmium solution from the tubes. The fibers should now appear black and have settled at the bottom of the tubes, making centrifugation unnecessary.
8. Wash the fibers three times in Milli-Q water for 10 min each.

9. Then wash the fibers in Milli-Q water using the microwave. Add fresh water and start the preinstalled "Water Rinse" program, which consists of the following settings: one cycle, 40 s, power 250 watts, no vacuum.
10. Replace the water in the tubes with fresh Milli-Q water and repeat step 9.
11. Replace the water in the tubes with 2% uranyl acetate aqueous solution (*see* **Note 5–7**).
12. **Do not close the tubes!** Under the chemical hood, transfer them to a vacuum chamber. Close the chamber, place it in the microwave processor, and connect it to the vacuum port using a rubber tube inside the instrument.
13. Start the preinstalled "Uranyl acetate staining" program, which consists of the following settings: 1 min of vacuum with microwaving (power: 150 watts), followed by 1 min of vacuum without microwaving. This cycle is repeated four times.
14. After completion, disconnect the vacuum chamber from the microwave and return it to the chemical hood.
15. Close the tubes and incubate them at room temperature in the dark for 2 h (*see* **Note 4**).
16. Without changing the solution in the tubes, repeat steps 12–14.
17. Transfer the tubes to the chemical hood and carefully remove the uranyl acetate.
18. Repeat washing steps 8–9.

3.3 Dehydration and Resin Embedding

1. After the final washing step, replace the water with 30% ethanol and incubate for 10 min. Then repeat this step sequentially with 50%, 70%, 90%, and 96% anhydrous ethanol, followed by anhydrous acetone (*see* **Note 8**).
2. Replace the anhydrous acetone with fresh one and incubate for 1 h.
3. Replace the acetone with a mixture of epoxy resin: acetone in the following ratios: 1:2, 1:1, and 2:1, incubating for 1 h at each step at room temperature (*see* **Note 2**).
4. After the final step, immerse the samples in 100% resin and leave them at room temperature overnight (*see* **Note 9**).
5. Transfer the samples to 60°C for 48 h to allow polymerization, then proceed with the routine protocol of ultrathin section preparation for TEM imaging [11].

4 Notes

1. Do not confuse the two chemicals! Use potassium ferrocyanide (pale yellow crystals), not potassium ferricyanide (bright orange-red crystals).
2. Dissociated fibers demonstrate good preservation of cellular structure and show no issues with sectioning after resin embedding. This allows us to conclude that the fixative and resin penetrate well and do not require microwave-assisted processing during these steps.
3. Ferrocyanide-reduced osmium staining is commonly used to enhance membrane contrast and preserve fine ultrastructural details. It is generally assumed that regular (non-reduced) osmium stains cell components more uniformly, whereas reduced osmium provides more selective and intense staining of membranes. However, the choice between reduced and regular osmium is not straightforward, as their effectiveness can vary depending on the sample type. For example, based on our experience with murine skeletal muscle, reduced osmium provides strong contrast in membrane structures, particularly the sarcoplasmic reticulum, while protein components such as actin and myosin are less prominent. In contrast, regular osmium yields strong contrast for protein structures, while the sarcoplasmic reticulum is barely visible. Meanwhile, we observed the opposite staining patterns in *Drosophila* muscles. Therefore, when beginning work with a new type of specimen, it is advisable to test both reduced and regular osmium staining to determine which is more suitable for your specific task. It is also possible to use both sequentially, with thiocarbohydrazide treatment in between. Additionally, testing both 1% and 2% osmium concentrations on your specific sample type is recom mended to determine the optimal conditions.
4. Microwave-assisted osmium tetroxide and uranyl acetate staining generally does not require additional incubation at room temperature between microwaving cycles; this step is mainly used to cool down the samples, which become heated during microwave treatment and only enhances contrast to a limited extent. If needed, the process can be accelerated by cooling the samples on ice between cycles.
5. When working with uranyl acetate, exposure to light should be minimized, as it is light-sensitive and can precipitate under prolonged light exposure.
6. Uranyl acetate is a radioactive compound and must be handled in accordance with institutional safety guidelines for radioactive materials. Appropriate protective equipment, designated

workspaces, and waste disposal procedures must be used at all times.

7. We use en bloc uranyl acetate staining as part of our conventional TEM protocol. Uranyl acetate staining can also be postponed until the grids stage if the safety regulation allows it.
8. The conventional dehydration protocol is currently under discussion. Some groups suggest starting dehydration at least at 50% or even 70% ethanol, based on the idea that lower ethanol concentrations have a greater potential to induce tissue damage. Therefore, 30% ethanol should be completely omitted.
9. When starting resin embedding, carefully replace the resin–acetone mixtures. Each replacement step should be performed slowly by gently removing the mixture and adding the new one along the wall of the tube to avoid disturbing the fibers settled at the bottom. Shaking is not allowed. If the fibers detach from the bottom and float, it becomes nearly impossible to reposition them, which complicates further work with the polymerized block. Centrifugation does not work in resin. We recommend using the same 1.5 mL plastic tubes in which the fibers were processed as molds for polymerization to avoid transferring the samples. Fill each tube with approximately 300–350 μL of resin and proceed with polymerization.

Acknowledgments

S.M. was supported by the Institute of Cytology and Genetics' project no. FWNR-2026-0024. O.A. acknowledges funding from the Abisch-Frenkel Foundation, the Minerva Foundation with funding from the Federal German Ministry for Education and Research, the Minna James Heineman Foundation, the Henry Chanoch Krenter Institute for Biomedical Imaging and Genomics, the Schwartz Reisman Collaborative Science Program, and the Yeda-Sela Center for Basic Research. O.A. is an incumbent of the Miriam Berman presidential development chair.

References

1. Goel AJ, Krauss RS (2019) Ex vivo visualization and analysis of the muscle stem cell niche. Methods Mol Biol 2002:39–50. https://doi.org/10.1007/7651_2018_177
2. Gineste C, Youhanna S, Vorrink SU et al (2022) Enzymatically dissociated muscle fibers display rapid dedifferentiation and impaired mitochondrial calcium control. iScience 25:105654. https://doi.org/10.1016/j.isci.2022.105654
3. Youhanna S, Bruton J, Jardemark K et al (2023) Calcium measurements in enzymatically dissociated or mechanically microdissected mouse primary skeletal muscle fibers. STAR Protoc 4:102260. https://doi.org/10.1016/j.xpro.2023.102260
4. Anderson JE, Wozniak AC, Mizunoya W (2012) Single muscle-fiber isolation and culture for cellular, molecular, pharmacological, and evolutionary

studies. In: DiMario J (ed) Myogenesis. Methods in molecular biology, vol 798. Humana Press, Totowa, NJ, pp 137–154. https://doi.org/10.1007/978-1-61779-343-1_6
5. Melaré R, Floriano RS, Gracia M et al (2016) Ultrastructural aspects of mouse nerve-muscle preparation exposed to Bothrops jararacussu and Bothrops bilineatus venoms and their toxins BthTX-I and Bbil-TX: unknown myotoxic effects. Microsc Res Tech 79:1082–1089. https://doi.org/10.1002/jemt.22748
6. Díaz-Castro F, Tuñón-Suárez M, Rivera P et al (2024) A single bout of resistance exercise triggers mitophagy, potentially involving the ejection of mitochondria in human skeletal muscle. Acta Physiol 240:e14203. https://doi.org/10.1111/apha.14203
7. Vonk LA, Esen O, Yuen M, Kirby TJ (2023) High-throughput contractile measurements of hydrogel-embedded intact mouse muscle fibers using an optics-based system. J Vis Exp 195:e65103. https://doi.org/10.3791/65103
8. Mayers CP (1970) Histological fixation by microwave heating. J Clin Pathol 23:273–275. https://doi.org/10.1136/jcp.23.3.273
9. Leong ASY, Sormunen RT (1998) Microwave procedures for electron microscopy and resin-embedded sections. Micron 29:397–409. https://doi.org/10.1016/S0968-4328(98)00018-3
10. Webster P (2014) Microwave-assisted processing and embedding for transmission electron microscopy. Methods Mol Biol 1117:21–37. https://doi.org/10.1007/978-1-62703-776-1_2
11. Kuo J (ed) (2007) Electron microscopy: methods and protocols. Methods in molecular biology, 2nd ed. Humana Press, Totowa, NJ. https://doi.org/10.1007/978-1-59745-294-6

Chapter 9

Analysis of Muscle Structure Using Fluorescently Tagged Proteins in *Drosophila melanogaster* Femur Muscles

Tiara Mulder, Mayar Ibrabah, Francesca Di Cara and Nicanor González-Morales

Abstract

The femur skeletal muscles of *Drosophila melanogaster* provide an excellent model for studying skeletal muscle development, function, physiology, and maintenance. Their key advantage over other adult skeletal muscles is the ability to image them with minimal dissection through the exoskeleton without compromising their structure, attachment sites, and overall morphology. Here, we present a general protocol for sample preparation and mounting of *Drosophila* legs that enables direct microscopy imaging of skeletal muscle structure. We provide representative images of the femur skeletal muscles under Apotome structured illumination microscopy and laser-scanning confocal microscopy. Finally, we highlight useful GFP or mCherry knock-in alleles of common skeletal muscle proteins that we have used successfully.

Keywords *Drosophila*, Skeletal muscles, Femur muscles, Live imaging, Sarcomere structure, Fluorescent proteins

1 Introduction

A sarcomere is the minimal contractile unit of skeletal muscles. It is bordered by two Z-discs, which serve as the boundaries of each sarcomere and anchor the thin filaments, primarily made up of the protein actin [1]. These thin filaments extend toward the center of the sarcomere, where they overlap with thick filaments composed of the protein myosin [1]. The thick filaments centers are at the M line, located at the center of the sarcomere. The arrangement of thick and thin filaments in the sarcomere is crucial for skeletal muscle contraction [1].

When a skeletal muscle contracts, the thin filaments slide past the thick filaments in a process known as the sliding filament theory. The myosin heads on the thick filaments bind to actin on the thin filaments and pull them toward the center of the sarcomere, shortening the sarcomere. This action is powered by ATP and regulated by calcium ions, which bind to troponin on the thin filaments, allowing

Joe Quadrilatero (ed.), *Myogenesis and Skeletal Muscle Biology*, Methods in Molecular Biology, vol. 3057,
https://doi.org/10.1007/978-1-0716-5396-8_9,

myosin to engage with actin. As the skeletal muscle contracts and the sarcomeres shorten, the Z-discs move closer together.

The general skeletal muscle structure is conserved between vertebrates and insects, making *Drosophila melanogaster* an invaluable model system for studying skeletal muscle biology and disease [2]. *Drosophila* has a powerful genetic toolkit, including the GAL4/UAS system, which enables precise manipulation of skeletal muscle-specific genes and visualization of fluorescently labeled proteins [3]. RNAi libraries allow for the depletion of virtually every gene in the genome [4]. Stock centers provide access to tens of thousands of transgenic lines, including insertional alleles and endogenously GFP-tagged genes [5–7]. Additionally, FlyBase is an exceptional database that compiles nearly all resources related to the *Drosophila* model [8]. In addition to the genetic tools, skeletal muscle differentiation and development are well characterized in *Drosophila*, myofibrillar proteins are highly conserved, and recent discoveries of skeletal muscle maintenance processes further establish flies as an ideal model for studying skeletal muscle biology [9–12]. Here, we present a simple method to dissect and visualize the femur structure using available fluorescent protein lines that avoids disturbing the femur muscle attachment and overall morphology of femur muscles.

1.1 Anatomy of the Drosophila *Leg*

The adult *Drosophila* leg is segmented, consisting of the coxa, trochanter, femur, tibia, and tarsus (Fig. 1). Males and females share these segmented sections, but males have sex combs on the first

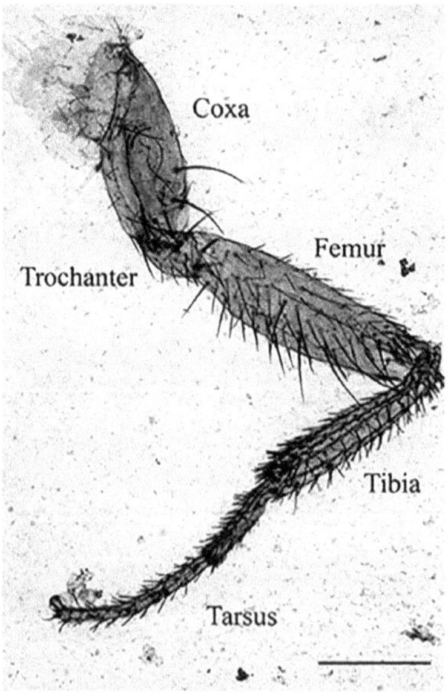

Fig. 1 Segmentation of the *Drosophila melanogaster*. Brightfield image showing the major segments of the *Drosophila* leg. The proximal coxa is attached to the thorax, followed by the trochanter, which connects to the femur. The distal segments include the tibia, bearing sensory bristles, and the tarsus, composed of several subsegments at the end with claws and adhesive pads (not visible here). Scale bar corresponds to 100 μm

pair of legs. Within each leg, there are 14 skeletal muscles made up of roughly 140 myofibrils, excluding the tarsus, which is devoid of skeletal muscles [13–16]. Each segment of skeletal muscles moves the subsequent segment, with the coxa skeletal muscle contributing to the trochanter's rotation and stabilization during walking and the femur skeletal muscles controlling the tibia [13, 14].

Within the femur, there are three distinct skeletal muscles: the tibia depressor muscle, tibia levator muscle, and tibia reductor muscle [14]. The tibia depressor muscle, which moves the tibia downward, has a sarcomere length of approximately 2.5 μm under relaxing conditions [14]. Opposing it is the tibia levator skeletal muscle, responsible for lifting the tibia, with a relaxed sarcomere length of about 3.5 μm. Positioned distally between the levator and depressor muscles is the tibia reductor muscle, which modulates tibial flexion and exhibits a relaxed sarcomere length of roughly 4.0 μm [14]. Internal tube-shaped tendons connect these skeletal muscles to the exoskeleton, facilitating force transmission during contraction [14–17]. Unlike the indirect flight skeletal muscles, which contract asynchronously via stretch activation, enabling multiple contractile cycles per neuronal input, leg skeletal muscles are synchronous, contracting in direct response to each motor neuron signal.

1.2 Fluorescent Markers for Imaging Drosophila Skeletal Muscles

There are many available *Drosophila* transgenic lines expressing skeletal muscle proteins tagged with fluorescent proteins, which allow visualizing different regions of the sarcomere. For example, the Z-disc can be visualized using markers such as α-actinin [18], Sls [19], and Zasp proteins [20–22], while the M-line can be visualized using Obscurin and Tropomodulin [23] (Figs. 2 and 3). For a summary of additional fluorescent proteins available for sarcomere imaging, refer to Table 1. Resilin is a protein found in the cuticle and lumen of the tracheal system and can serves as an effective secondary marker when imaging leg muscles due to its strong autofluorescence in the blue spectral range [24]. Resilin autofluorescence allows visualization of cuticular structures and tracheal morphology, providing spatial context and aiding in the interpretation of skeletal muscle organization relative to external anatomy (Fig. 3). In conjunction with resilin, other cellular organelles and structures can be visualized depending on the research questions, using, for example, nuclear or mitochondrial proteins (*see* **Note 1**) [23]. Myosin and actin can be visualized using either endogenous fluorescently tagged proteins or antibodies via dissection (see Table 1 and Fig. 4).

1.3 Muscle Measurements

Muscle measurements can be performed in ImageJ using a variety of tools in response to a variety of research questions. The line tool allows for accurate measurement of sarcomere length as well as the width of individual sarcomeres. Fluorescence intensity can be quantified by using the freehand tool to select an area, for example, the Z-disc, and pressing Ctrl + M, which provides a rough estimate of protein amount if tagged with fluorescence in the selected area.

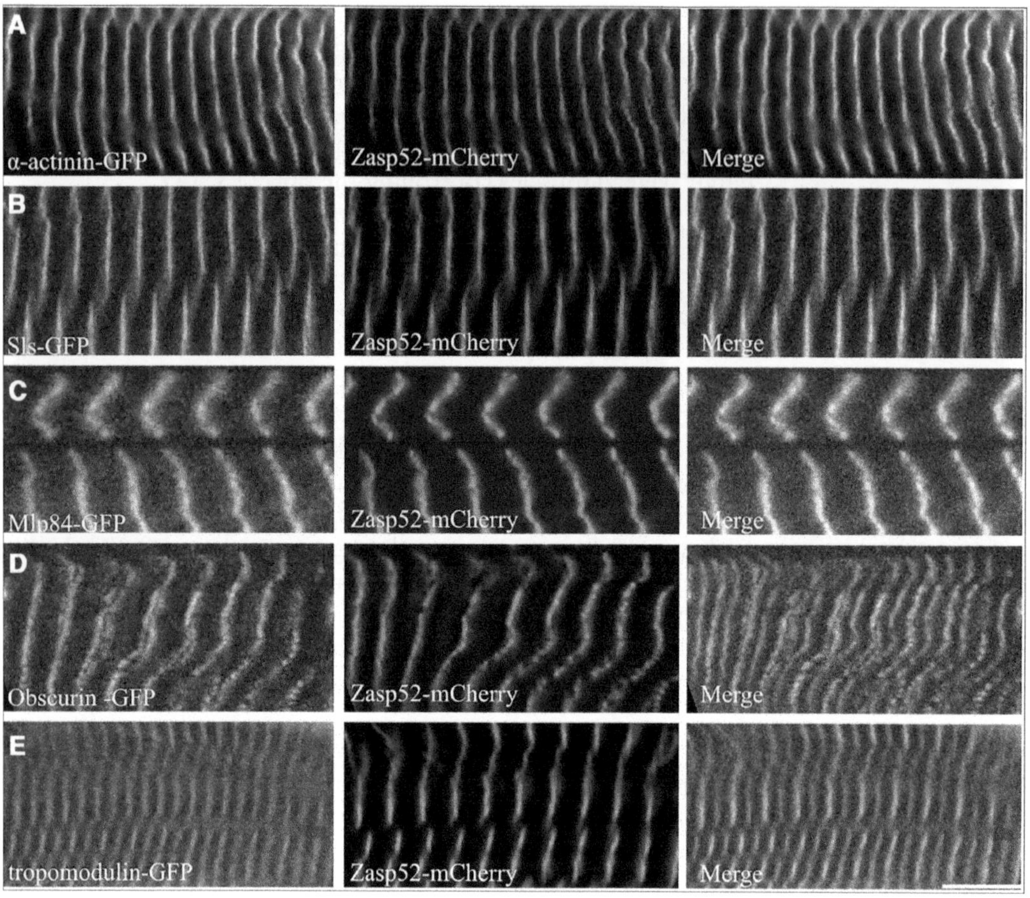

Fig. 2 Example of Apotome images using different fluorescent proteins. (**A**) Depressor muscles expressing α-actinin-GFP in green and Zasp52-mCherry in magenta. (**B**) Depressor muscles expressing Sls-GFP in green and Zasp52-mCherry in magenta. (**C**) Depressor muscles expressing Mlp-GFP in green and Zasp52-mCherry in magenta. (**D**) Depressor muscles expressing Obscurin-GFP in green and Zasp52-mCherry in magenta. (**E**) Depressor muscles expressing Tropomodulin-GFP in green and Zasp52-mCherry in magenta. In panels (**A**)–(**C**), all proteins are localized to the Z-disc as expected. In panels (**D**) and (**E**), the alternating Z-disc and M-line signals (as seen by alternating colors of green and magenta) are apparent in all depressor muscles observed. The levator and depressor muscles have very similar sarcomere organization. Scale bar corresponds to 2 μm

The freehand selection tool can also be used to measure the area of the muscle(s), providing insight into overall fiber size. By drawing a line across a myofibril and pressing Ctrl + K, a plot profile is generated that allows for comparison of signal strength across the Z-disc and M-line or other areas of interest (Fig. 5). As shown in Fig. 5, the alternating bands of the Z-disc and M-line can be visualized. In addition, ImageJ can be used to measure fiber diameter, sarcomere orientation/angle, and spacing regularity using Fourier analysis or line profiles. ImageJ also offers an extensive suite of plugins that can aid in the quantification of colocalization between fluorescent

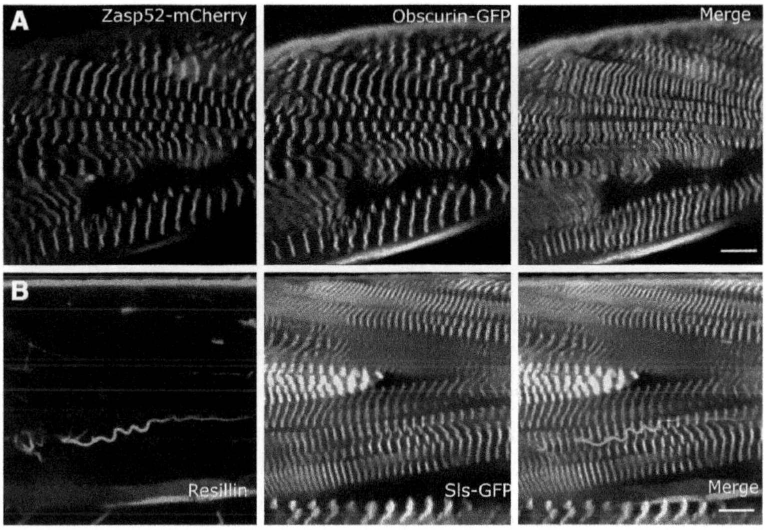

Fig. 3 Example of confocal images using different fluorescent proteins. (**A**) Femur skeletal muscles expressing Obscurin-GFP in green and Zasp52-mCherry in magenta. The alternating Z-disc and M-line signals are apparent in all skeletal muscles observed. (**B**) Femur skeletal muscle expressing Sls-GFP in green and using the autofluorescence of resilin in magenta (405 ex/ 440 em). Resilin marks the lumen of the tracheal system and the cuticle. It is useful to observe the overall morphology of the leg and the tracheal system. Scale bar corresponds to 10 μm

Table 1
Selected useful fluorescent proteins to analyze the femur muscles

Stock	Flybase ID	References
filamin$^{WT.mGFP6}$	FBal0387772	[25]
Zasp52^{G00189}	FBal0148383	[26]
Zasp52^{MI02988}	FBal0276969	[27]
Zasp66^{ZCL0663}	FBal0150936	[26]
Zasp67$^{fTRG-1384\ sfGFP}$	FBal0339687	[5, 22]
α-Actinin^{CC01961}	FBal0211606	[18, 28]
Obscurin$^{fTRG01046.sfGFP}$	FBtp0129013	[5, 23]
Mlp84B^{GFP}	FBal0240776	[29]
Sls^{ZCL2144}	FBal0230218	[30]
UAS-LifeAct-Ruby	FBti0143328	[31]
Mhc^{YD0783}	FBti0100292	[32]
Tmod$^{MI07502-GFSTF}$	FBti0178532	[6]

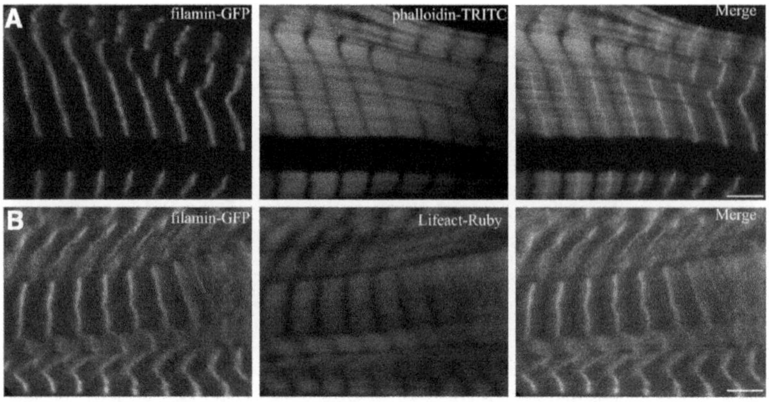

Fig. 4 Representative Apotome images of actin filaments in the levator muscles. (**A**) Actin staining in levator muscles using TRITC-phalloidin in red, with filamin-GFP in green as a Z-disc marker. (**B**) Lifeact-Ruby in magenta in the levator muscle marks the thin filaments in a fixed sample, with filamin-GFP in green as a Z-disc marker. Scale bar corresponds to 2 μm

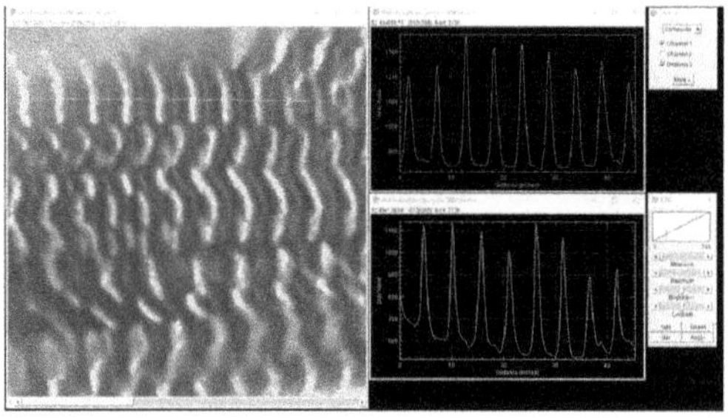

Fig. 5 Visualization of Zasp52 and Obscurin localization in *Drosophila* femur muscles. Confocal image of *Drosophila* femur muscle fibers showing alternating bands of Zasp52-mCherry in magenta at the Z-Disc and Obscurin-GFP in green. A line was drawn across multiple sarcomeres to generate intensity profiles for each protein channel. The corresponding fluorescence intensity plots (right) reveal the periodic banding pattern of Z-discs (Zasp52-mCherry in magenta) and M-lines (Obscurin-GFP in green), highlighting their distinct localization within the sarcomere

proteins, intensity distribution within defined regions, and even 3D reconstructions of muscle volume when Z-stacks are available. Therefore, ImageJ provides a comprehensive toolkit for analyzing skeletal muscle structure, organization, and protein localization and can be used across dissected, live, or fixed leg preparations (Fig. 6).

2 Materials

1. Fly stocks (stocks used will vary depending on the experiment).
2. Conventional needles Precision Glide 25G, 5/8 in.
3. Cover Glass No. 1½, 22 × 30 mm.

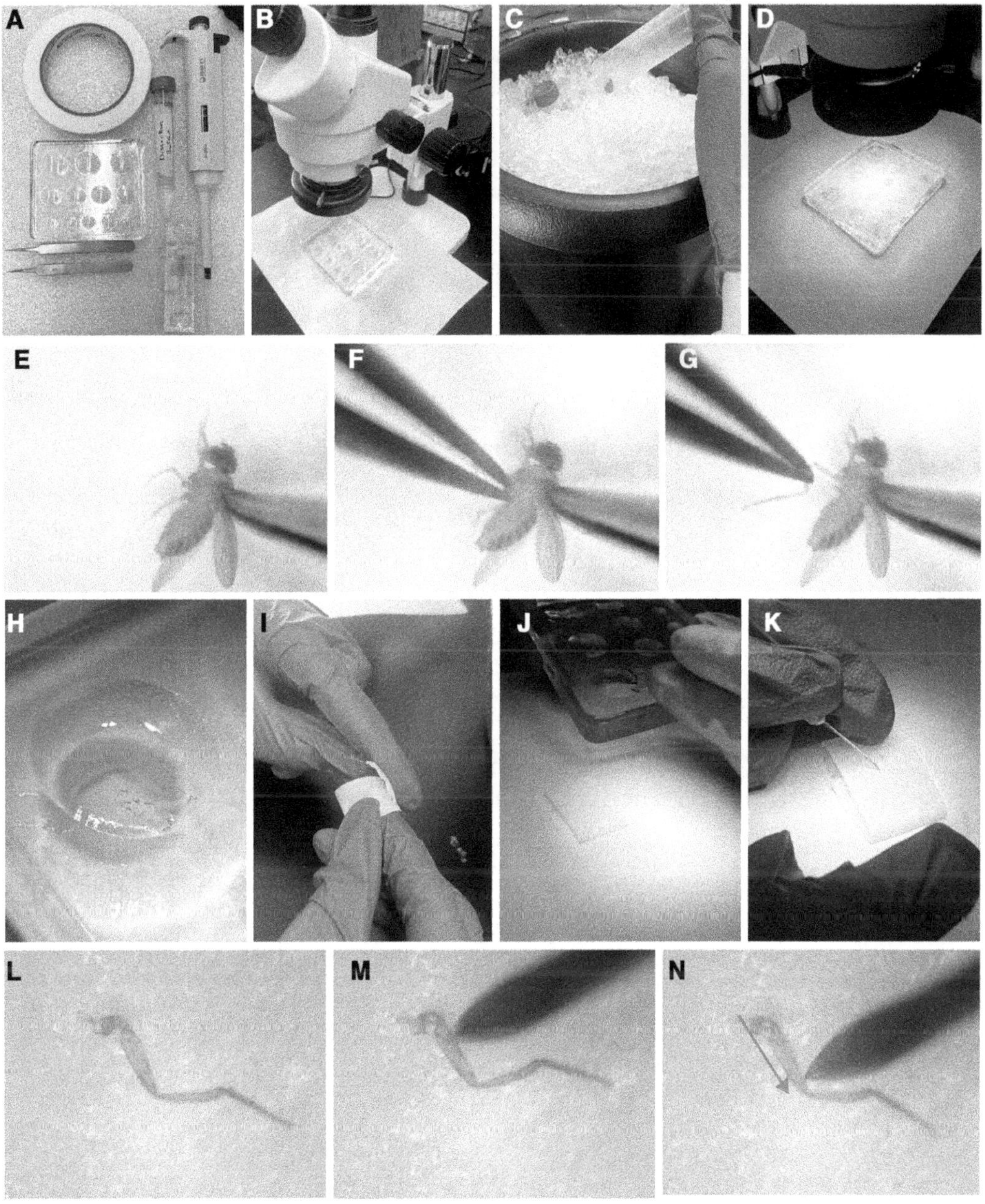

Fig. 6 Leg dissection protocol. (**A**) Equipment required for dissection. (**B**) Microscope and well-plate setup. (**C**) Place *Drosophila* into an empty vial and then place the vial on ice until the flies are anesthetized. (**D**) Using a pipette, add a small amount of leg buffer into an empty well of the plate, and place the anesthetized fly into this well (*see* **Note 3**). (**E**) Using forceps, gently grasp the thorax and rotate the fly onto its side to expose the legs. (**F**) With the other forceps, gently grasp the leg at the proximal coxa, near the thoracic junction. (**G**) Apply a gentle pulling force until the leg detaches. (**H**) Leave the freshly detached legs in the well plate with the leg buffer and remove all other segments. (**I**) Using a piece of tape, form a circular loop by attaching the adhesive side of one end to the non-adhesive side of the opposite end, creating a ring-shaped tape structure. Place this ring down onto the piece of paper. (**J**). Using forceps, pick up the legs from the tarsus and transfer them onto the tape. (**K** and **L**) Using the needle, gently press down on the legs to adhere to the tape. (**M**) Place the needle at the top of the femur, right below the trochanter. (**N**) Drag the needle longitudinally downward gently yet firmly down the femur, repeat four to five times to create microtears in the cuticle

4. Microscope slides.
5. Phalloidin–tetramethyl rhodamine B isothiocyanate (TRITC) (Sigma-Aldrich, catalog number: P1951).
6. Mowiol 4-88 mounting media (Sigma-Aldrich, catalog number: 81381).
7. Adenosine triphosphate (ATP).
8. Glycerol.
9. 8% paraformaldehyde.
10. Sodium phosphate dibasic anhydrous (Na_2HPO_4).
11. Sodium chloride (NaCl).
12. Potassium chloride (KCl).
13. Potassium phosphate monobasic (KH_2PO_4).
14. Tween 20.
15. Ethyleneglycol-*bis*(β-aminoethyl)-N,N,N′,N′-tetraacetic Acid (EGTA).
16. Leg dissection buffer (50 mL total): Add 1.0 mL of 1 M sodium phosphate buffer, pH 7.0 (final concentration: 20 mM), 0.5 mL of 0.5 M $MgCl_2$ (final concentration: 5 mM), 2.5 mL of 0.1 M EGTA (final concentration: 5 mM), 3.75 mL of 0.1 M ATP stock solution (final concentration: 7.5 mM), 50 μL of Tween-20, and 25 mL of glycerol to a 50 mL graduated cylinder or glass beaker. Add ddH_2O to a final volume of 50 mL. Mix thoroughly. Aliquot ~10 mL into each of ten small Falcon tubes. Store at −20°C.
17. Sodium phosphate buffer (NaPi) 1 M (100 mL total): Add 12.8 g of sodium phosphate monobasic and 3.2 g of sodium phosphate dibasic to ~80 mL of ddH_2O in a 100 mL graduated cylinder or glass beaker. Adjust the pH to 7.4. Add ddH_2O to a final volume of 100 mL. Autoclave. Store at room temperature.
18. EGTA solution 0.5 M (50 mL total): Add 9.5 g of EGTA to ~40 mL of ddH_2O in a 50 mL graduated cylinder or glass beaker. Mix until dissolved. Add ddH_2O to a final volume of 50 mL. Store at room temperature.
19. Phosphate-buffered solution (PBS) 10× (1 L total): Add 25 g of Na_2HPO_4, 80 g of NaCl, 2 g of KCl, and 2 g of KH_2PO_4 to a 1 L graduated cylinder or glass beaker. Add ddH_2O to a final volume of 1 L. Mix thoroughly. Store in the refrigerator.
20. Phosphate-buffered saline with Tween-20 (PBST) 1× (1 L total): Add 100 mL of 10× PBS to a 1 L graduated cylinder or glass beaker. Add 900 mL of ddH_2O. Add 1 mL of Tween-20. Mix thoroughly. Store in the refrigerator.

21. Confocal microscopy: Thorlabs laser-scanning confocal microscope: Use a 60× Plan Apo 1.42 NA Nikon oil immersion objective. Acquire image stacks to visualize the levator and depressor muscles in the femur. Use mCherry (excitation/emission: 587/610 nm) and GFP (excitation/emission: 565/654 nm) as fluorescent markers. Include the DAPI channel (excitation/emission: 358/461 nm) to visualize resilin. Maintain identical acquisition settings across all samples for comparability.
22. Structured illumination microscopy: Zeiss Apotome 3: Use a 60× oil objective. Select the appropriate LED channels for fluorophore excitation. Acquire image stacks that allow visualization of the levator and depressor muscles in the femur. Use mCherry (excitation/emission: 587/610 nm), GFP (excitation/emission: 565/654 nm), and DAPI (excitation/emission: 358/461 nm, for resilin) as fluorescent markers. Maintain identical acquisition settings across all samples for comparability.
23. Immersion oil Immersol 518F (Zeiss, catalog number: 433802-9010-000).
24. Type F immersion liquid (Lecica Microsystems, catalog number: 11 513 859).
25. Two high-precision dissection tweezers (Towot SA-14 forceps or similar).
26. Dissecting scope (Olympus, catalog number: SZX7).
27. Ice packs.
28. Bucket of ice.
29. Masking tape.
30. Sheet of white paper.

3 Methods

3.1 Leg Dissection for Fixed or Live Imaging

1. Place an ice pack under the microscope stage, cover it with a sheet of white paper, and then position the well plate on top to keep samples cool during dissection.
2. Anesthetize adult *Drosophila* by chilling them on ice (Fig. 5C).
3. As flies are being anesthetized, place one drop of leg dissection buffer into an empty well plate (*see* **Note 2**).
4. Once anesthetized, transfer flies (any sex, any adult age) into the well plate containing chilled leg buffer (Fig. 6D).
5. With the forceps, gently grasp the thorax and turn the fly to expose the legs (Fig. 6E).

6. With forceps, gently grasp the leg at the proximal coxa, close to the thoracic junction (Fig. 6F).
7. Detach the leg by applying a gentle pulling force until separation occurs, leaving the legs inside the leg buffer and removing all other segments (Fig. 6G, H).
8. For live imaging, transfer dissected legs grasping by the tarsus onto a microscope slide with a small droplet of fresh leg buffer (*see* **Note 4**).
9. Carefully place a coverslip over the sample to maintain tissue hydration and viability.
10. For fixed imaging, incubate the dissected legs in 4% paraformaldehyde (PFA) diluted in 1× PBST for 1 h at room temperature on a shaker.
11. After 2 h, wash fixed samples twice in 1× PBS for 10 min per rinse on a shaker.
12. Mount the legs Mowiol 4-88 mounting media and allow the mounting media to set at 4°C for at least 24 h.
13. Proceed with imaging using an Apotome structured illumination microscope or a laser-scanning confocal microscope.

3.2 Leg Dissection for Phalloidin or Antibody Staining

1. Repeat steps 1–7 from Sect. 3.1.
2. Using a piece of tape, form a circular loop by attaching the adhesive side of one end to the non-adhesive side of the opposite end, creating a ring-shaped tape structure (Fig. 6I).
3. Gently remove legs from the leg buffer grasping at the tarsus and place them onto the adhesive side of the prepared tape (Fig. 6J).
4. Using the needle, gently push down on the legs to adhere them securely to the tape (Fig. 6K, L).
5. Place a small drop of leg buffer onto the legs.
6. Using the tip of the fine needle, carefully position the needle directly under the trochanter (Fig. 6M).
7. Gently yet firmly drag the needle longitudinally downward along the length of the femur, creating microtears (Fig. 6K–N).
8. Repeat four to five times.
9. Incubate the dissected legs in 8% PFA diluted in 1× PBST to make 4% PFA for 1 h on a shaker.
10. Rinse the fixed legs in PBST for 5 min on a shaker.
11. Perform two additional rinses in PBS, each lasting 5 min each on a shaker.

12. Incubate legs overnight at 4°C on a shaker in PBST containing phalloidin TRITC in a 1:1000 ratio.
13. Following incubation, rinse twice in PBST for 5 min each on a shaker.
14. If desired, mount the legs using Mowiol 4-88 mounting media and allow the mounting media to set at 4°C for at least 24 h.
15. Proceed with imaging using the desired microscopy and microscope techniques.

Without the leg dissection buffer, the levator has a reduction in sarcomere size within 15 min of imaging. (A) Levator skeletal muscles dissected with leg buffer, with the Z-disc marked with Zasp 52, have a sarcomere size 3.5 mm, and the mCherry is localized to the Z-disc. (B) Levators skeletal muscles dissected with PBS, with the Z-disc marked with Zasp52, have a reduced sarcomere size of 2.5 mm, and the mCherry becomes diffused across the leg, with a less crisp signal.

4 Notes

1. Note that due to the thickness and low permeability of the cuticle, antibody-based staining requires to open a hole(s) in the cuticle and, even then, is considerably less effective. Actin filaments can be imaged using fluorescent proteins such as Lifeact-Ruby or staining with fluorescent conjugated phalloidin (Fig. 4). Lifeact-Ruby allows marking specific skeletal muscles using the UAS-Gal4 system live imaging. However, the signal intensity is not as strong as with phalloidin. One disadvantage of using phalloidin is that it can only be used for fixed samples (Fig. 4).
2. Leg dissection buffer should be kept on ice as much as possible to prevent ATP degradation.
3. In addition to optimizing microscopy settings, researchers must also carefully consider sample preparation. Based on observations in our lab, dissecting legs in PBS followed by live imaging or fixation in PBS results in sarcomere shortening (Fig. 7). Therefore, dissecting directly in the leg dissection buffer is required (Fig. 7).
4. Live imaging should commence within 15 min post-dissection due to ATP degradation.

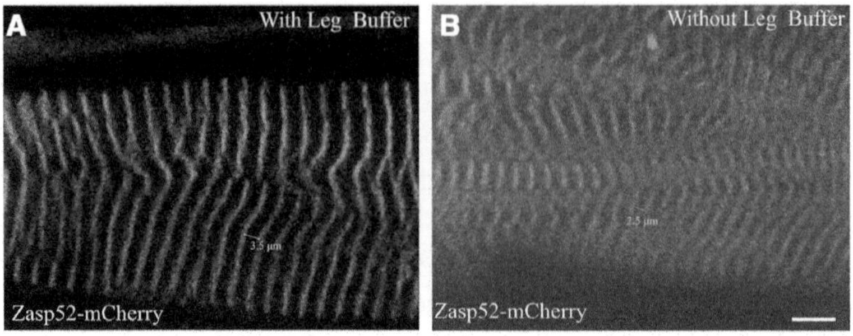

Fig. 7 An ATP-containing buffer is required for proper femur muscle imaging. Confocal images of the levator muscle with Zasp52-mCherry in magenta under two buffer conditions. (**A**) Levator muscle imaged in the leg dissection buffer containing ATP shows a clear sarcomere structure. (**B**) Levator muscles imaged in standard PBS buffer exhibit increased background noise and shorter sarcomeres, indicative of a contracted state

References

1. Schöck F, González-Morales N (2022) The insect perspective on Z-disc structure and biology. J Cell Sci 135. https://doi.org/10.1242/jcs.260179
2. Boukhatmi H (2021) Drosophila, an integrative model to study the features of muscle stem cells in development and regeneration. Cells 10. https://doi.org/10.3390/cells10082112
3. Brand AH, Perrimon N (1993) Targeted gene expression as a means of altering cell fates and generating dominant phenotypes. Development 118:401. https://doi.org/10.1242/dev.118.2.401
4. Hu Y, Comjean A, Rodiger J et al (2021) FlyRNAi.org—the database of the Drosophila RNAi screening center and transgenic RNAi project: 2021 update. Nucleic Acids Res 49:D908–D915. https://doi.org/10.1093/nar/gkaa936
5. Sarov M, Barz C, Jambor H et al (2016) A genome-wide resource for the analysis of protein localisation in Drosophila. eLife 5:e12068. https://doi.org/10.7554/eLife.12068
6. Nagarkar-Jaiswal S, Lee PT, Campbell ME et al (2015) A library of MiMICs allows tagging of genes and reversible, spatial and temporal knockdown of proteins in Drosophila. eLife 4:e05338. https://doi.org/10.7554/eLife.05338
7. Buszczak M, Paterno S, Lighthouse D et al (2007) The Carnegie protein trap library: a versatile tool for Drosophila developmental studies. Genetics 175:1505–1531. https://doi.org/10.1534/genetics.106.065961
8. Gramates LS, Agapite J, Attrill H et al (2022) FlyBase: a guided tour of highlighted features. Genetics 220:iyac035. https://doi.org/10.1093/genetics/iyac035
9. Mulder T, Johnson J, González-Morales N (2025) The filamins of Drosophila. Genome 68:1–11. https://doi.org/10.1139/gen-2024-0159
10. Boukhatmi H, Bray S (2018) A population of adult satellite-like cells in Drosophila is maintained through a switch in RNA-isoforms. eLife 7:e35954. https://doi.org/10.7554/eLife.35954
11. Chaturvedi D, Reichert H, Gunage RD, VijayRaghavan K (2017) Identification and functional characterization of muscle satellite cells in Drosophila. eLife 6:e30107. https://doi.org/10.7554/eLife.30107
12. Fisher LAB, Carriquí-Madroñal B, Mulder T et al (2024) Filamin protects myofibrils from contractile damage through changes in its mechanosensory region. PLoS Genet 20:e1011101. https://doi.org/10.1371/journal.pgen.1011101
13. Laurichesse Q, Soler C (2020) Muscle development: a view from adult myogenesis in Drosophila. Semin Cell Dev Biol 104:39–50. https://doi.org/10.1016/j.semcdb.2020.02.009
14. Soler C, Daczewska M, Da Ponte JP et al (2004) Coordinated development of muscles and tendons of the Drosophila leg. Development 131:6041–6051. https://doi.org/10.1242/dev.01527
15. Baek M, Mann RS (2009) Lineage and birth date specify motor neuron targeting and dendritic architecture in adult Drosophila.

J Neurosci 29:4939–4947. https://doi.org/10.1523/JNEUROSCI.1585-09.2009
16. Guan W, Venkatasubramanian L, Baek M et al (2018) Visualize Drosophila leg motor neuron axons through the adult cuticle. J Vis Exp. 2018 Oct 30;(140). https://doi.org/10.3791/58365
17. Moucaud B, Prince E, Jagla K, Soler C (2023) Developmental origin of tendon diversity in Drosophila melanogaster. Front Physiol 14:1176148. https://doi.org/10.3389/fphys.2023.1176148
18. González-Morales N, Xiao YS, Schilling MA et al (2019) Myofibril diameter is set by a finely tuned mechanism of protein oligomerization in Drosophila. eLife 8:e50496. https://doi.org/10.7554/eLife.50496
19. González-Morales N, Holenka TK, Schöck F (2017) Filamin actin-binding and titin-binding fulfill distinct functions in Z-disc cohesion. PLoS Genet 13:e1006880. https://doi.org/10.1371/journal.pgen.1006880
20. Liao KA, González-Morales N, Schöck F (2016) Zasp52, a core Z-disc protein in Drosophila indirect flight muscles, interacts with α-actinin via an extended PDZ domain. PLoS Genet 12:e1006400. https://doi.org/10.1371/journal.pgen.1006400
21. Liao KA, González-Morales N, Schöck F (2020) Characterizing the actin-binding ability of Zasp52 and its contribution to myofibril assembly. PLoS One 15:e0232137. https://doi.org/10.1371/journal.pone.0232137
22. González-Morales N, Marsh TW, Katzemich A et al (2019) Different evolutionary trajectories of two insect specific paralogous proteins involved in stabilizing muscle myofibrils. Genetics 212:267–281. https://doi.org/10.1534/genetics.119.302217
23. González-Morales N, Marescal O, Szikora S et al (2023) The oxoglutarate dehydrogenase complex is involved in myofibril growth and Z-disc assembly in Drosophila. J Cell Sci 136:jcs260717. https://doi.org/10.1242/jcs.260717
24. Lerch S, Zuber R, Gehring N et al (2020) Resilin matrix distribution, variability and function in Drosophila. BMC Biol 18: 195. https://doi.org/10.1186/s12915-020-00902-4
25. Huelsmann S, Rintanen N, Sethi R et al (2016) Evidence for the mechanosensor function of filamin in tissue development. Sci Rep 6: 32798. https://doi.org/10.1038/srep32798
26. Katzemich A, Liao KA, Czerniecki S, Schöck F (2013) Alp/Enigma family proteins cooperate in Z-disc formation and myofibril assembly. PLoS Genet 9:e1003342. https://doi.org/10.1371/journal.pgen.1003342
27. Xiao YS, Schöck F, González-Morales N (2017) Rapid IFM dissection for visualizing fluorescently tagged sarcomeric proteins. Bio Protoc 7:e2606. https://doi.org/10.21769/bioprotoc.2606
28. Vakaloglou KM, Chountala M, Zervas CG (2012) Functional analysis of parvin and different modes of IPP-complex assembly at integrin sites during Drosophila development. J Cell Sci 125:482–494. https://doi.org/10.1242/jcs.102384
29. Clark KA, Bland JM, Beckerle MC (2007) The Drosophila muscle skeletal muscle LIM protein, Mlp84B, cooperates with D-titin to maintain muscle structural integrity. J Cell Sci 120:2066–2077. https://doi.org/10.1242/jcs.000695
30. Orfanos Z, Leonard K, Elliott C et al (2015) Sallimus and the dynamics of sarcomere assembly in Drosophila flight muscles. J Mol Biol 427:2151–2168. https://doi.org/10.1016/j.jmb.2015.04.003
31. Hatan M, Shinder V, Israeli D et al (2011) The Drosophila blood-brain barrier is maintained by GPCR-dependent dynamic actin structures. J Cell Biol 192:307–319. https://doi.org/10.1083/jcb.201007095
32. Hudson AM, Petrella LN, Tanaka AJ, Cooley L (2008) Mononuclear muscle cells in Drosophila ovaries revealed by GFP protein traps. Dev Biol 314:329–340. https://doi.org/10.1016/j.ydbio.2007.11.029

Chapter 10

High-Resolution Skeletal Muscle Fiber Typing Using Four-Color Immunofluorescence Microscopy

Erin Allison, Mackenzie Q. Graham, Fasih A. Rahman and Joe Quadrilatero

Abstract

Skeletal muscle is composed of heterogeneous fiber types defined by distinct myosin heavy chain (MYH) isoforms. Accurate fiber type identification is essential for studying skeletal muscle development, plasticity, disease, and intervention responses. Traditional fiber type classification techniques are limited by either lack of specificity or the inability to simultaneously detect all fiber types in a single section. Here, we describe a novel four-channel immunofluorescent staining protocol that enables simultaneous visualization and quantification of type I, IIA, IIX, and IIB fibers within a single cross-section of rodent skeletal muscle. This high-resolution method improves throughput and eliminates the need for serial sectioning but also allows for the identification of hybrid fibers. Moreover, membrane staining can be applied to simultaneously perform cross-sectional area analysis. This approach enhances the precision and efficiency of skeletal muscle phenotyping in both physiological and pathological contexts by capturing the full spectrum of skeletal muscle fibers in a single image, thus providing a powerful tool for investigating fiber type-specific development and adaptation.

Keywords Skeletal muscle, Fiber typing, Myosin heavy chain, Immunofluorescence, Hybrid fibers, Skeletal muscle plasticity, Skeletal muscle development

1 Introduction

Skeletal muscle is a heterogeneous tissue composed of fiber types that differ in their contractile and metabolic characteristics [1–3]. This heterogeneity enables skeletal muscle to perform a wide range of physiological functions, from maintaining posture to generating rapid movements. Fiber-type specification begins during development, where early myogenic precursors express embryonic and neonatal myosin heavy chain (MYH) isoforms, which are progressively replaced by adult isoforms in a temporally regulated and muscle-specific manner [1]. The phenotypic traits of individual skeletal muscle fibers

Erin Allison and Mackenzie Q. Graham contributed equally.

Joe Quadrilatero (ed.), *Myogenesis and Skeletal Muscle Biology*, Methods in Molecular Biology, vol. 3057,
https://doi.org/10.1007/978-1-0716-5396-8_10,

are shaped by both intrinsic factors like gene profile and extrinsic factors including neural input, mechanical loading, and circulating signals [1]. As a result, skeletal muscle exhibits remarkable plasticity in response to physiological stimuli (e.g., exercise) and pathological conditions (e.g., myopathies). These adaptations may be beneficial resulting in increased muscle mass or improved metabolic efficiency, or detrimental leading to impaired function, muscle wasting, or metabolic dysregulation [2, 4]. One of the most widely adopted frameworks for characterizing skeletal muscle fiber heterogeneity is based on the expression of specific MYH isoforms. In humans, three adult MYH isoforms exist including MYHI (MYH7), MYHIIA (MYH2), and MYHIIX (MYH1), representing type I, type IIA, and type IIX fibers, respectively. In certain species, such as rodents, an additional MHCIIB (MYH4) isoform is present, representing type IIB fibers [1, 3, 5]. These fiber types can differ significantly in contractile kinetics, fatigue resistance, mitochondrial content, and enzymatic activity [1]. As such, accurate classification and quantification of fiber types is critical for investigating skeletal muscle function, plasticity, disease pathogenesis, and responses to interventions. A variety of techniques have historically been used to classify skeletal muscle fiber types, each with its own strengths and limitations. Early approaches relied on the differential sensitivity of myosin ATPase activity to pre-incubation at varying pH levels, enabling separation of fiber types [6–8]. More recently, immunohistochemical labeling with antibodies targeting specific MYH isoforms has become the preferred standard due to its improved specificity [9, 10]. While these techniques have been instrumental in advancing our understanding of skeletal muscle biology, many of the earlier approaches now lack the precision and throughput needed for current applications, particularly related to fiber type-specific analyses. Our group previously developed a three-channel immunofluorescent staining method to enable rapid quantification of fiber types [3]. However, this approach leaves one fiber type unstained or requires analysis of multiple serial sections, which is time-consuming, particularly in mixed rodent skeletal muscles consisting of all four fiber types. In this chapter, we introduce a novel four-channel immunofluorescent staining protocol capable of simultaneously detecting four distinct MYH isoforms within a single muscle section. In addition, this approach provides a sensitive approach to identify hybrid fibers.

2 Materials

1. Biological skeletal muscle sample.
2. 1× phosphate-buffered saline (PBS).
3. Goat serum.

4. Optimal cutting temperature (OCT) compound.
5. Isopentane.
6. Liquid nitrogen.
7. Aluminum can.
8. Cork for freezing tissue.
9. Large forceps for retrieving tissue.
10. Cryo-storage container (mini dewar).
11. Vectabond reagent (SP-1800-7; Vector Laboratories).
12. 100% acetone.
13. Clear nail polish.
14. Glass microscope slides.
15. 24 mm × 24 mm #1 coverslips.
16. Cryostats blades.
17. Fine paint brush.
18. Aluminum foil.
19. Light-protected humidified chamber.
20. Hydrophobic pen.
21. Fluoromount-G mounting medium or similar.
22. Pipette tips (various sizes).
23. Refrigerator.
24. -80°C freezer.
25. Cryostat.
26. Light microscope.
27. Widefield/Confocal microscope (e.g., Zeiss LSM 800).
28. Blocking solution: 10% goat serum in 1× PBS.
29. Primary antibody solution: Dilute the following primary antibodies in blocking solution immediately before use:
    a. type I (BA-F8, mouse IgG2b, 1:25).
    b. type IIA (SC-71, mouse IgG1, 1:500).
    c. type IIX [6H1-(Kb9L)2, rabbit IgG, 1:20].
    d. type IIB (BF-F3, mouse IgM, 1:50).
    e. dystrophin (MANDYS1, mouse IgG2a, 1:500, *see* **Note 1**).
30. Secondary antibody solution: Dilute secondary antibodies in blocking solution at 1:500 immediately before use (*see* **Note 2**):
    a. DyLight 405–conjugated anti-mouse IgG2b.
    b. Alexa Fluor 488–conjugated anti-mouse IgG1.

c. Alexa Fluor 555–conjugated anti-mouse IgM.
d. Alexa Fluor 555–conjugated anti-mouse IgG2a.
e. Alexa Fluor 647–conjugated anti-rabbit IgG.

3 Methods

3.1 Sample Collection and Freezing

1. Dissect skeletal muscle (e.g., gastrocnemius, soleus, or EDL) promptly following euthanasia.
2. Gently blot/remove excess blood and connective tissue.
3. Embed muscle in optimal cutting temperature (OCT) compound on a labeled piece of cork (*see* **Note 3**).
4. Snap-freeze the embedded tissue by submerging cork in a metal can filled with liquid isopentane pre-chilled in liquid nitrogen (~ −160°C).
5. Once fully frozen (typically within 10–30 s), transfer the OCT block to a cryostorage container and store at −80°C until sectioning.

3.2 Vectabond Slides

To improve tissue adhesion, slides are pre-treated with Vectabond reagent.

1. Submerge clean glass slides in 100% acetone for 5 min.
2. Transfer slides to a solution of 50:1 acetone:Vectabond reagent for 5 min.
3. Remove slides and air-dry in a fume hood.
4. Store treated slides in a clean, dry container at room temperature.

3.3 Cryosectioning

1. Equilibrate frozen skeletal muscle samples in the cryostat (−20°C) for 15–30 min prior to sectioning (*see* **Note 4**).
2. Mount the OCT block containing the skeletal muscle onto the specimen holder using a small amount of fresh OCT.
3. Cut transverse sections at a thickness of 10 μm (*see* **Note 5**).
4. Collect sections onto Vectabond treated glass slides (*see* **Note 6**).
5. Air dry slides at room temperature for 10–15 min and either proceed immediately with staining or store wrapped lightly in aluminum foil at −80°C for future use.

3.4 Immunofluorescent Staining

1. Bring slides to room temperature for 15 min and circumscribe tissue section with a hydrophobic pen.

2. Incubate sections in blocking solution for 1 h at room temperature.
3. Remove blocking solution from slides.
4. Apply primary antibody solution and incubate for up to 2 h at room temperature or overnight at 4°C in a humidified chamber.
5. Wash slides 3 × 5 min in 1× PBS.
6. Apply secondary antibody solution and incubate for 1 h at room temperature in a humidified chamber protected from light.
7. Wash slides 3 × 5 min in 1× PBS.
8. Mount coverslips using an antifade mounting medium (e.g., Fluoromount-G). Seal edges with clear nail polish and store slides at 4°C in the dark until imaging (*see* **Note 7**).

3.5 Imaging

1. Allow stained slides to equilibrate to room temperature and inspect coverslips for proper sealing prior to imaging (*see* **Note 8**).
2. Place slides on the stage and locate the region of interest at 10× or 20× magnification.
3. Select appropriate laser lines (e.g., 405, 488, 561, and 640 nm lasers using the Zeiss LSM 800) to excite each fluorophore.
4. Optimize laser intensity and detector gain.
5. Record exposure/power and gain settings. Apply the same parameters to all samples in each experiment.
6. Export individual and merged channel files in 16-bit TIFF format for analyses.

3.6 Fiber Type Distribution Analysis

1. Open the merged immunofluorescence image in FIJI/ImageJ. Ensure that all fluorophores (e.g., DyLight 405, Alexa Fluor 488, 555, and 647) are visible in the composite image, allowing fiber types to be identified by their unique fluorescence signatures (Fig. 1; *see* **Note 9**).
2. Navigate to Plugins > Analyze > Cell Counter > Cell Counter and click "Initialize." This will open a duplicate of your image in the Cell Counter window.
3. In the Cell Counter panel, select a different counter label to correspond to each fiber type:
   a. Label 1 = Type I fibers (e.g., DyLight 405).
   b. Label 2 = Type IIA fibers (e.g., Alexa Fluor 488).
   c. Label 3 = Type IIX fibers (e.g., Alexa Fluor 647).
   d. Label 4 = Type IIB fibers (e.g., Alexa Fluor 555).
   e. Additional labels can be added for hybrid fibers.

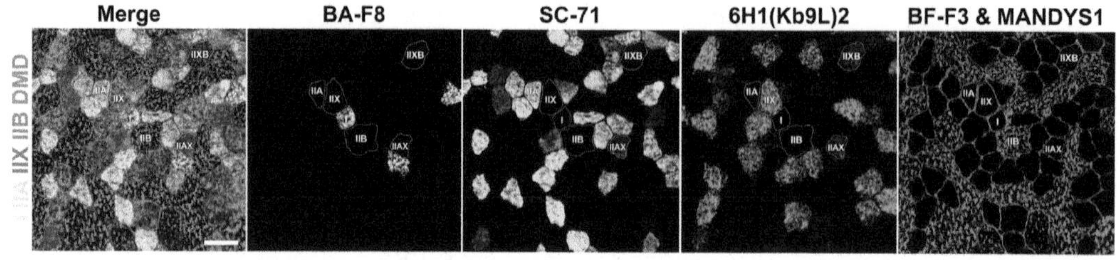

Fig. 1 Representative four-channel immunofluorescent staining of myosin heavy chain (MYH) isoforms in mouse red gastrocnemius muscle. Images include a merged composite and individual channels highlighting type I fibers (BA-F8; blue), type IIA (SC-71; green), type IIX (6H1(Kb9L)2; purple), and type IIB (BF-F3; red). Hybrid fibers co-expressing multiple isoforms (e.g., type IIAX and IIXB) are shown. Dystrophin (MANDYS1) is labeled in red to delineate sarcolemmal borders. Scale bar = 50 μm

4. Manually click the center of each fiber using the appropriate counter type. Move systematically across the image (e.g., left to right, top to bottom) to avoid double-counting or omissions. The running tally for each fiber type will be displayed in the Cell Counter panel (Fig. 2; *see* **Note 10**).
5. To correct a mistake, click "Delete Mode" and then click the incorrect marker. Once all fibers are counted, click "Results" in the Cell Counter panel.
6. This will generate a summary table showing the count for each fiber type. To save your counts: Click "Save Markers" to export an .xml file of marker positions for record-keeping or reloading.
7. Export the Results Table as a .csv file for statistical analysis. Calculate percent fiber type distribution:

$$\begin{aligned} &\textit{Fiber Type Distribution}\ (\%) \\ &= \frac{\textit{Number of Specific Fiber Type}}{\textit{Total Number of Fibers Counted}} * 100 \end{aligned}$$

3.7 Cross-Sectional Area (CSA) Analysis

1. Open the merged TIFF image in FIJI/ImageJ. Ensure that the sarcolemma (e.g., dystrophin or laminin) channel provides clear fiber borders for tracing.
2. Make sure Analyze > Set Measurements includes Area and Display Label.
3. Set the scale to convert pixel dimensions to microns (Analyze > Set Scale and enter the known pixel-to-micron ratio provided by the microscope software). Set "Unit of Length" to micron and check "Global" to apply it to all images in the session. This ensures area measurements will be reported in μm² rather than pixels.

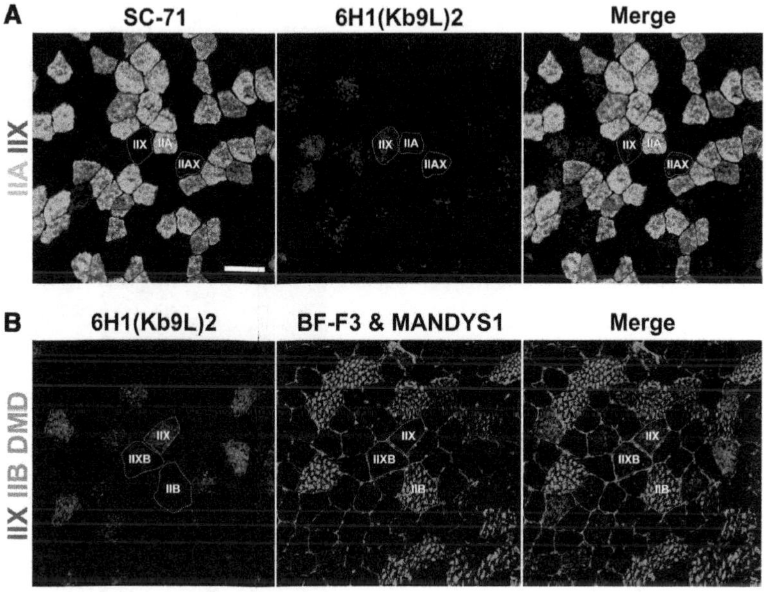

Fig. 2 Validation of novel antibody cocktail for detecting hybrid skeletal muscle fibers in mouse red gastrocnemius. (**A**) Representative images showing type IIA (SC-71; green) and type IIX (6H1(Kb9L)2; purple) labeling, along with the merged image. A hybrid type IIAX fiber co-expressing both isoforms (green+purple) is identified. (**B**) Representative images showing type IIX (6H1(Kb9L)2; purple) and type IIB (BF-F3; red) labeling, with corresponding merged image. A hybrid type IIXB fiber co-expressing both isoforms (purple+red) is identified. Dystrophin (MANDYS1) is also labeled in red to delineate sarcolemmal borders. Scale bar = 50 μm

4. Use the Freehand Selection Tool or Polygon Tool to manually trace around the inner edge of the sarcolemma for a fiber expressing a specific MYH marker (*see* **Note 11**).
5. Once traced, press "M" or go to Analyze > Measure to record the CSA. After measuring, press "Delete" or "Backspace" to cut out the fiber from the image. This provides a clear visual cue that the fiber has already been analyzed and helps avoid duplicate counting.
6. Continue this process systematically for one fiber type at a time (e.g., measure ~30 type I fibers, then move on to type IIA, etc.), working from top to bottom or left to right across the image.
7. For hybrid fibers, apply the same tracing and deletion approach, ensuring both fluorophores are co-expressed before assigning fiber type.
8. Aim to measure approximately 30 fibers per fiber type, evenly distributed across the entire cross-section to avoid regional bias. After completion, export the Results Table as a .csv file

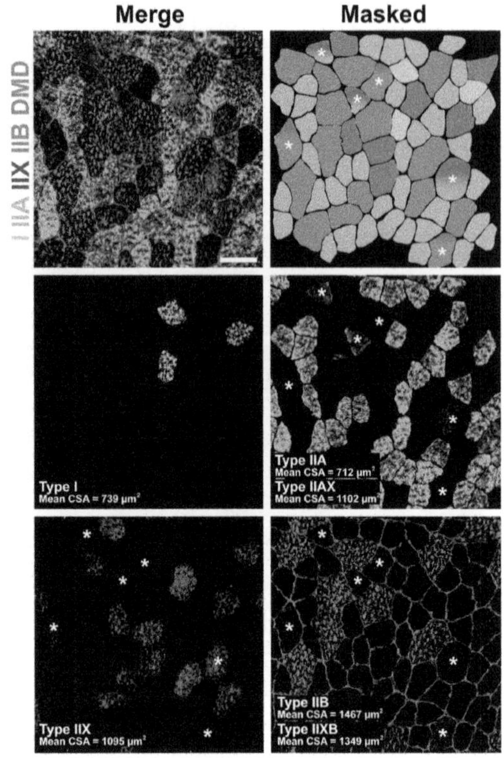

Fig. 3 Quantification of four-channel immunofluorescent staining of myosin heavy chain (MYH) isoforms in mouse red gastrocnemius muscle. Images include a merged composite and masked image. Individual channels showing type I fibers (BA-F8; blue), type IIA (SC-71; green), type IIX (6H1(Kb9L)2; purple), and type IIB (BF-F3; red). Hybrid fibers co-expressing multiple isoforms (e.g., type IIAX (green/purple) and IIXB (purple/red)) are denoted with an asterisk (*). Dystrophin (DMD; MANDYS1) is included in red to delineate sarcolemmal borders. Quantification of mean cross-sectional area (CSA) for each fiber type is presented in μm². Scale bar = 50 μm

(File > Save As > Results) for further analysis. Optionally, save the edited image (with removed fibers) for documentation and verification that all intended fibers were measured (Fig. 3).

4 Notes

1. Dystrophin is included in the antibody cocktail to outline the sarcolemma. Although it has a distinct antibody isotype (IgG2a vs. IgM), it shares the same fluorochrome as the type IIB antibody and is therefore detected using the same channel/parameters. Although this results in identical fluorescence labeling, dystrophin localizes exclusively to the

membrane and does not interfere with the interpretation of fiber-type staining. It can be excluded from analysis if needed.

2. Alternative fluorochromes may be used depending on experimental needs; however, they must be compatible with the isotype of the primary antibody to ensure selective and accurate detection. Matching the fluorochrome-conjugated secondary antibody to the correct antibody subclass (e.g., IgG1, IgG2b, IgM) is essential to avoid cross-reactivity and preserve signal specificity.
3. Use only the minimal amount of OCT compound necessary to support the tissue during sectioning, as excess OCT can interfere with subsequent staining or imaging. Before embedding tissue, ensure that a solid frozen layer of isopentane is visible at the base of the aluminum can; this indicates that the isopentane has reached a sufficiently low temperature for rapid tissue freezing.
4. If the sample has not equilibrated to the cryostat temperature, it may fracture or snap during sectioning due to thermal shock. Allow the tissue to reach thermal equilibrium within the cryostat chamber before sectioning. For skeletal muscle, a temperature between −18°C and −22°C is typically optimal, with −20°C as a standard starting point.
5. As each section comes off the blade, use an anti-roll plate or a fine brush to gently flatten it. This prevents folding and helps maintain proper orientation.
6. Use a light microscope to verify that sections are oriented transversely. If not, adjust the angle of the sample within the chuck and continue sectioning accordingly.
7. For high-resolution imaging, ensure that the mounting medium is free of air bubbles and that tissue sections are clean and free of debris.
8. Although it is best practice to image slides the day after mounting to allow the mounting medium and nail polish to fully dry, slides remain suitable for imaging for several days if stored in a light-protected container at 4°C.
9. To confirm fluorescence specificity and minimize spectral overlap, antibodies can be tested individually to verify their emission profiles. This control step helps ensure accurate channel separation and avoids misidentification of fiber types due to bleed-through or cross-reactivity.
10. Muscle fiber types can be identified based on the presence or absence of MYH isoform-specific fluorescence signals. Pure fibers express a single MYH isoform and appear in only one fluorescent channel. For example, type IIA fibers appear only green, while type IIX fibers appear only purple. Hybrid fibers

co-express two MYH isoforms and are identified by overlapping fluorescence signals, such as green and purple for type IIAX or red and purple for type IIXB. Careful inspection of merged images is essential to accurately distinguish true hybrid fibers from background fluorescence.

11. When performing CSA analysis, do not include fibers that appear elongated, distorted, or irregular in shape, as these are likely obliquely sectioned and can skew the results.

Acknowledgments

Funding was provided to J.Q. by the Natural Science and Engineering Research Council of Canada (NSERC; RGPIN 258590). F.A.R. was supported by a NSERC CGS-D scholarship. M.Q.G. was supported by a NSERC CGRS-M scholarship.

References

1. Schiaffino S, Reggiani C (2011) Fiber types in mammalian skeletal muscles. Physiol Rev 91:1447–1531
2. Rahman FA, Baechler BL, Quadrilatero J (2024) Key considerations for investigating and interpreting autophagy in skeletal muscle. Autophagy pp. 1–12
3. Bloemberg D, Quadrilatero J (2012) Rapid determination of myosin heavy chain expression in rat, mouse, and human skeletal muscle using multicolor immunofluorescence analysis. PLoS One 7:e35273
4. Schiaffino S, Reggiani C, Akimoto T, et al (2021) Molecular Mechanisms of Skeletal Muscle Hypertrophy. J Neuromuscul Dis 8:169–183
5. Smerdu V, Karsch-Mizrachi I, Campione M, et al (1994) Type IIx myosin heavy chain transcripts are expressed in type IIb fibers of human skeletal muscle. Am J Physiol 267:C1723–C1728
6. Brooke MH, Kaiser KK (1970) Muscle fiber types: how many and what kind?. Arch Neurol 23:369–379
7. Delp MD, Duan C (1996) Composition and size of type I, IIA, IID/X, and IIB fibers and citrate synthase activity of rat muscle. J Appl Physiol 80:261–270
8. Guth L, Samaha FJ (1970) Procedure for the histochemical demonstration of actomyosin ATPase. Exp Neurol 28:365–367
9. Schiaffino S, Gorza L, Sartore S et al. (1989) Three myosin heavy chain isoforms in type 2 skeletal muscle fibres. J Muscle Res Cell Motil 10:197–205
10. Lucas CA, Kang LHD, Hoh JFY (2000) Monospecific antibodies against the three mammalian fast limb myosin heavy chains. Biochem Biophys Res Commun 272:303–308

Chapter 11

Chronic, Low-Frequency Electrical Pulse Stimulation of Primary Human Skeletal Muscle Cells

Eili Tranheim Kase and Nataša Nikolić

Abstract

Electrical pulse stimulation (EPS) represents a useful in vitro model of exercise in skeletal muscle cells. By modifying stimulation patterns, e.g., voltage, frequency, pulse duration, and total stimulation time, it is possible to generate models that simulate different types of in vivo exercise under controlled conditions. Currently, there is a wide variability among existing protocols. Here, we describe a chronic, low-frequency protocol for EPS of human myotubes in 6-well plates using single bipolar pulses of 2 ms, 1 Hz at 10 V for 24 h.

Keywords Human skeletal muscle, Myotubes, Contractions, Electrical pulse stimulation

1 Introduction

Physical activity is strongly associated with improved health and well-being, playing an important role in the battle against many conditions, including diabetes, cardiovascular diseases, and obesity [1–3]. Skeletal muscle is a highly adaptable organ, with a fascinating ability to adjust and remodel in response to exercise. Molecular mechanisms and biochemical pathways underlying these substantial adaptations in muscle, but also in other organs, are, however, still poorly understood. Motor neuron activation of muscle cells, comprising both mechanical and electrical signals, regulates growth and differentiation processes [4]. These signals can be replaced or mimicked by electrical pulse stimulation (EPS) in cell culture, leading to increased contractions, accelerated sarcomere assembly, and changes in metabolic profile of the cells [5–11]. It has therefore been proposed that EPS can be used to mimic muscle contractions both in vivo and ex vivo, representing a useful model to study exercise-induced responses under controlled conditions [11, 12].

Exercise has many beneficial health effects, which may vary depending on intensity and duration of exercise. Ever since the initial

Joe Quadrilatero (ed.), *Myogenesis and Skeletal Muscle Biology*, Methods in Molecular Biology, vol. 3057,
https://doi.org/10.1007/978-1-0716-5396-8_11,

development of EPS, it has been desirable to identify the type of exercise in vitro in a similar manner to in vivo. Thus, over the last decade, numerous studies using various EPS protocols have been developed, aiming to replicate different exercise modalities (chronic/acute, endurance/resistance, low/high intensity) [11, 12]. Existing EPS protocols involve many factors, including electric potentials (volts), frequency (Hz), pulse duration (ms), and stimulation duration time. There is, however, a great amount of heterogeneity among current EPS protocols, even for the same type of exercise, making it rather challenging to define the type of exercise in vitro based exclusively on the applied EPS parameters. Several biological markers, including AMPK (5' adenosine monophosphate-activated protein kinase), Akt (protein kinase B), IL-6 (interleukin-6), PGC-1α (peroxisome proliferator-activated receptor-gamma coactivator α), and glucose uptake, have been proposed as parameters of exercise in vitro [12]. These biological markers have been shown to increase in EPS-treated cells across studies, following patterns of in vivo exercise. Importantly, the observed effects appear to be correlated with duration of the stimulation, with a trend for a reduction in the effect of EPS with increasing duration [12].

There are mainly two types of electrical pulse stimulators used in the majority of currently published studies involving EPS: commercially available and custom-made. When it comes to cell types, EPS is frequently applied to either cell lines or primary cells. Specifically, C2C12 cell line (a mouse myoblast cell line) and L6 cell line (a rat myoblast cell line) have been successfully used with EPS [6, 13, 14]. Among primary cells, the majority of studies have been performed in skeletal muscle cells from humans [7, 8, 10], but there are also studies performed in primary cells obtained from mice [15]. In studies involving EPS in human primary skeletal muscle cells, muscle biopsies have been obtained from different donor groups (healthy, lean, obese, diabetic) [7–10, 16–19]. Cell culture systems derived from human muscle biopsies have been used in research for decades. Given their genetic background, metabolic, and biochemical properties, they represent the most valuable model to study metabolic properties of adult skeletal muscles under controlled conditions [20].

Differentiated human skeletal muscle cells (myotubes) are quiescent in cultures and do not typically contract spontaneously unless being innervated or electrically stimulated [11, 21]. Some of the observed effects of chronic, low-frequency EPS of human myotubes, reported by ours and several other research groups, include reorganization of the cytoskeleton, de novo formation of sarcomeric structures, visible contractions, and profound metabolic adaptations reflected in improved oxidative capacity and increased insulin sensitivity, as well as identification of novel, contraction-induced myokines [7, 8, 16, 22].

In this chapter, we describe an EPS protocol for chronic, low-frequency stimulation of human myotubes using single bipolar pulses of 2 ms, 1 Hz, and 10 V for a duration of 24 h using a commercially available electrical pulse stimulator. For methods describing acute, high-frequency EPS using a custom-made electrical pulse stimulator, as well as for low-frequency EPS of longer duration (48 h), we refer to our previously published method chapter [23].

2 Materials

1. Cells: EPS is performed on adherent, differentiated myotubes. We use human myotubes obtained from human satellite cells. A bank of myoblasts is established from human muscle biopsy samples obtained with informed consent and according to regulations of national ethical committees, following a modified previously described method [24]. Briefly, muscle tissue is placed in a Petri dish containing 5 ml DMEM and carefully minced and dissected. The tissue is then dissociated by three successive treatments of 20 min each in 0.10% Trypsin/EDTA at room temperature. Dissociated cells are centrifuged at 550 g for 7 min and resuspended in fully supplemented skeletal muscle cell basal medium (PromoCell) using collagen-coated (0.005%) 25 cm^2 flasks. Cells are cultured in an incubator at 37°C and 5% CO_2, with media change every 2–3 days, until confluence is obtained. Cells are stored in liquid nitrogen.
2. Proliferation medium: DMEM- Glutamax™ (5.5 mM glucose) supplemented with 10% heat-inactivated FBS, 50 U/ml penicillin, 50 µg/ml streptomycin, 1.25 µg/ml amphotericin B, 50 mg/ml gentamicin, 0.05% BSA, 0.39 µg/ml dexamethasone, 10 µg/ml hEGF, and 25 mM HEPES.
3. Differentiation medium: DMEM- Glutamax™ (5.5 mM glucose) supplemented with 2% heat-inactivated FBS, 50 U/ml penicillin, 50 µg/ml streptomycin, 1.25 µg/ml amphotericin B, 50 mg/ml gentamicin, 25 pM insulin, and 25 mM HEPES.
4. Culturing plates: Corning CellBIND six-well plates (Corning Life-Sciences, Schiphol-Rijk, The Netherlands).
5. Pulse generator and C-Dish electrodes. Chronic, low-frequency EPS is performed using a commercially available IonOptix C-Pace EP cell culture stimulator (Fig. 1A) with C-Dishes (Fig. 1B), both from IonOptix, Dublin, Ireland.

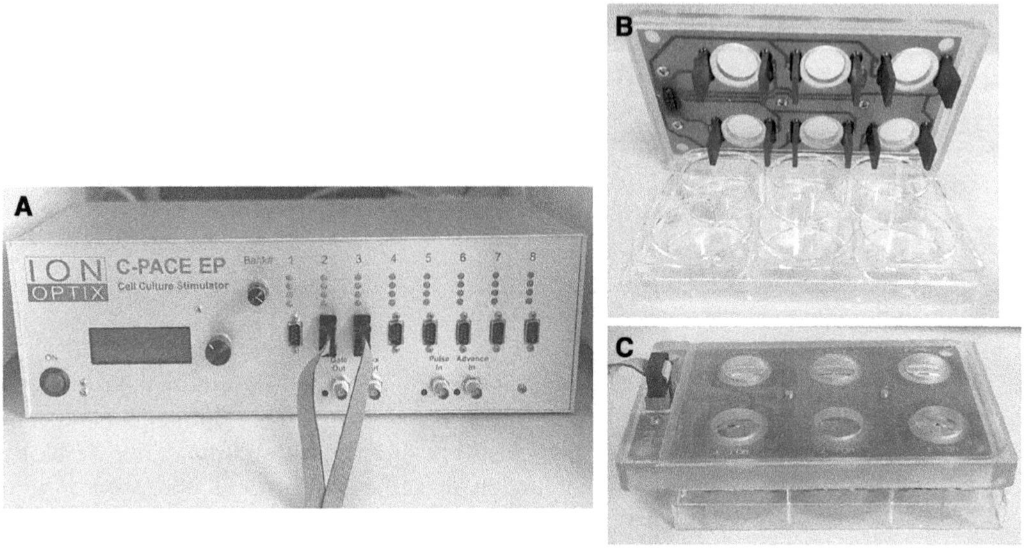

Fig. 1 Pulse generator and C-Dish used for chronic, low-frequency EPS. (**A**) IonOptix C-Pace pulse generator. (**B**) C-Dish with carbon electrodes. (**C**) C-Dish fitted on top of six-well cell culturing plate

3 Methods

All work involving cells and C-Dish electrodes is carried out in a cell culture safety hood using aseptic working technique. Cells are cultured in a humidified 5% CO_2 atmosphere at 37°C.

3.1 Establishment of Differentiated Human Skeletal Muscle Cells (Myotubes)

1. Human myoblasts are taken out from liquid nitrogen, thawed, and seeded at a density of 100,000 cells/well in a Corning CellBIND (*see* **Note 1**) 6-well plate, using 1–2 ml/well (*see* **Note 2**) of proliferation medium. Medium is changed every 2–3 days.
2. Differentiation of myoblasts into myotubes is induced at approximately 80% confluence by switching to differentiation medium, which is changed every 2–3 days. Cells are differentiated for 7 days.
3. Prior to EPS, myotubes should be attached to the surface of the culturing plate and fully differentiated. Inspect the cells frequently under a microscope, both during culturing and EPS (*see* **Note 3**).

3.2 Electrical Pulse Stimulation (EPS)

1. Chronic, low-frequency EPS is applied on day 6 of the 7-day differentiation period.
2. Sterilize the C-Dishes immediately before use by spraying with 70% ethanol and allowing them to air dry completely.

3. Fresh, warm (37°C) differentiation medium should be added to the cells before EPS is started.
4. Fit the C-Dish on top of the culturing plate so that the C-Dish electrodes descend into the culturing media. The cell culture plate lid will fit the groove on top of the C-Dish (Fig. 1C).
5. Transfer assemblies of culturing plates and C-Dishes into the incubator.
6. Connect assemblies to the pulse generator (Fig. 1A) with appropriate cables, run cables through the door opening of the incubator, and apply stimulation protocol as anticipated. For chronic, low-frequency stimulation, use single, bipolar pulses of 2 ms, 10 V, and 1 Hz continuously for the last 24 h of a 7-day differentiation period (*see* **Note 4**).
7. After the end of stimulation, cells are harvested, and C-Dishes are cleaned and sterilized as described in the following section (*see* **Note 5**).

3.3 Cleaning and Sterilization of C-Dishes

After each use, clean and sterilize electrodes following the next procedure (*see* **Note 6**):

1. After the application ends, soak the entire C-Dish in distilled water for a couple of days with daily changing of water.
2. Use a soft toothbrush to scrub the surface of the electrodes under rinsing distilled water to avoid accumulation of absorbed salts (*see* **Note 7**).
3. After cleaning, C-Dishes must be sterilized.
4. Sterilization is performed using either of the following procedures (*see* **Note 8**):
   a) Disinfection with 70% ethanol, followed by placing the C-Dish under UV germicidal lamp in a cell culture hood for 15 min.
   b) Heating the C-Dish in a drying oven at 100°C for 3 h.
5. Cleaned and sterilized C-dishes are placed on top of sterile 6-well culturing plates, with the culture plate lid fitted on top, wrapped in aluminum foil, and stored at room temperature.

4 Notes

1. Cells should be attached to the surface of the cell culturing plate. Corning CellBIND plates are frequently used for EPS. However, human myotubes originating from different donors may respond differently to the same EPS parameters. Coating

and/or optimaization of stimulation parameters may be necessary to avoid cell detachment. The present protocol has been successfully applied to cells originating from healthy, lean, obese (with and without T2D), and donors with spinal cord injury. When introducing a new type of cells to the existing EPS-protocol, inspect the cells carefully to detect detachments. For cells that detach under EPS when cultured on Corning CellBIND plates, EPS has been successfully performed by coating plates with extracellular matrix (ECM) or Matrigel®. For details, see our previously published method chapter [23].

2. Electrodes must be immersed in liquid during EPS, and the volume of the medium is adjusted accordingly. Volume of 2.2 ml is sufficient when Corning CellBIND six-well plates and compatible six-well C-Dishes are used. The volume required will vary depending on the length of the electrodes and the types of plates. Always check that the electrodes are immersed in media and adjust volume if necessary.
3. When visually inspecting myotubes for contractions during EPS, avoid vigorous movements. Visualizing contractions in human myotubes may be challenging. Not all cells contract at the same time, so make sure to take time when visually inspecting cells for contractions.
4. For chronic, low-frequency EPS, voltages from 8 to 30 V may be used at different durations. For a longer duration of stimulation (48 h), we refer to our previously published chapter [23]. Preferably, voltage should not be higher than 30 V for the described protocol designed around a six-well assembly, as higher voltage increases the risk of overheating.
5. The culturing medium does not need to be changed when the duration of EPS exposure is no longer than 24 h.
6. Carbon electrodes absorb salts, electrolysis by-products, and proteins from the media. Toxic chemicals may leach from the electrodes into the media, causing nonspecific effects or cell death. Performing the entire cleaning and sterilization procedure is highly recommended for a successful EPS experiment.
7. The toothbrush may turn black once the rubbing of the electrodes is finished. Carbon is a solid material, and this should not be a matter of concern.
8. Cleaning and sterilization (procedure "a" in our hands) of C-Dishes is essential to prevent contamination. Using an autoclave for sterilization is not recommended [23]. Importantly, all work involving cells and C-Dishes is carried out in a cell culture safety hood using aseptic working technique.

References

1. Colberg SR, Sigal RJ, Yardley JE et al (2016) Physical activity/exercise and diabetes: a position statement of the American Diabetes association. Diabetes Care 39(11):2065–2079. https://doi.org/10.2337/dc16-1728
2. Wewege MA, Thom JM, Rye KA et al (2018) Aerobic, resistance or combined training: a systematic review and meta-analysis of exercise to reduce cardiovascular risk in adults with metabolic syndrome. Atherosclerosis 274:162–171. https://doi.org/10.1016/j.atherosclerosis.2018.05.002
3. Flouris AD, Bouziotas C, Christodoulos AD et al (2008) Longitudinal preventive-screening cutoffs for metabolic syndrome in adolescents. Int J Obes (Lond) 32(10):1506–1512. https://doi.org/10.1038/ijo.2008.142
4. Song Y, Soto J, Chen B et al (2020) Cell engineering: biophysical regulation of the nucleus. Biomaterials 234:119743. https://doi.org/10.1016/j.biomaterials.2019.119743
5. Thelen MH, Simonides WS, van Hardeveld C (1997) Electrical stimulation of C2C12 myotubes induces contractions and represses thyroid-hormone-dependent transcription of the fast-type sarcoplasmic-reticulum Ca2+-ATPase gene. Biochem J 321(Pt 3):845–848
6. Fujita H, Nedachi T, Kanzaki M (2007) Accelerated de novo sarcomere assembly by electric pulse stimulation in C2C12 myotubes. Exp Cell Res 313(9):1853–1865. https://doi.org/10.1016/j.yexcr.2007.03.002
7. Raschke S, Eckardt K, Bjorklund Holven K et al (2013) Identification and validation of novel contraction-regulated myokines released from primary human skeletal muscle cells. PLoS One 8(4):e62008. https://doi.org/10.1371/journal.pone.0062008
8. Nikolic N, Bakke SS, Kase ET et al (2012) Electrical pulse stimulation of cultured human skeletal muscle cells as an in vitro model of exercise. PLoS One 7(3):e33203. https://doi.org/10.1371/journal.pone.0033203
9. Mars T, Mis K, Meznaric M et al (2021) Innervation and electrical pulse stimulation—in vitro effects on human skeletal muscle cells. Appl Physiol Nutr Metab 46(4):299–308. https://doi.org/10.1139/apnm-2019-0575
10. Lovsletten NG, Rustan AC, Laurens C et al Primary defects in lipid handling and resistance to exercise in myotubes from obese donors with and without type 2 diabetes. Appl Physiol Nutr Metab. 2019. 45(2):169–179. https://doi.org/10.1139/apnm-2019-0265.
11. Nikolic N, Gorgens SW, Thoresen GH et al (2017) Electrical pulse stimulation of cultured skeletal muscle cells as a model for *in vitro* exercise—possibilities and limitations. Acta Physiol (Oxf) 220(3):310–331. https://doi.org/10.1111/apha.12830
12. Nintou E, Karligiotou E, Vliora M et al Characteristics of the protocols used in electrical pulse stimulation of cultured cells for mimicking in vivo exercise: a systematic review, meta-analysis, and meta-regression. Int J Mol Sci. 2022;23(21):13446. https://doi.org/10.3390/ijms232113446
13. Nedachi T, Fujita H, Kanzaki M (2008) Contractile C2C12 myotube model for studying exercise-inducible responses in skeletal muscle. Am J Physiol Endocrinol Metab 295(5):E1191–E1204. https://doi.org/10.1152/ajpendo.90280.2008
14. Nedachi T, Hatakeyama H, Kono T et al (2009) Characterization of contraction-inducible CXC chemokines and their roles in C2C12 myocytes. Am J Physiol Endocrinol Metab 297(4):E866–E878. https://doi.org/10.1152/ajpendo.00104.2009
15. Bosutti A, Bernareggi A, Massaria G et al (2019) A "noisy" electrical stimulation protocol favors muscle regeneration *in vitro* through release of endogenous ATP. Exp Cell Res 381(1):121–128. https://doi.org/10.1016/j.yexcr.2019.05.012
16. Feng YZ, Nikolic N, Bakke SS et al (2015) Myotubes from lean and severely obese subjects with and without type 2 diabetes respond differently to an *in vitro* model of exercise. Am J Physiol Cell Physiol 308(7):C548–C556. https://doi.org/10.1152/ajpcell.00314.2014
17. Aas V, Torbla S, Andersen MH et al (2002) Electrical stimulation improves insulin responses in a human skeletal muscle cell model of hyperglycemia. Ann N Y Acad Sci 967:506–515
18. Mengeste AM, Nikolic N, Dalmao Fernandez A et al (2022) Insight into the metabolic adaptations of electrically pulse-stimulated human myotubes using global analysis of the transcriptome and proteome. Front Physiol 13:928195. https://doi.org/10.3389/fphys.2022.928195
19. Jan V, Mis K, Nikolic N et al (2021) Effect of differentiation, de novo innervation, and electrical pulse stimulation on mRNA and protein expression of Na+,K+-ATPase, FXYD1, and FXYD5 in cultured human skeletal muscle cells. PLoS One 16(2):e0247377. https://doi.org/10.1371/journal.pone.0247377
20. Aas V, Bakke SS, Feng YZ et al (2013) Are cultured human myotubes far from home? Cell Tissue Res. 354:671. https://doi.org/10.1007/s00441-013-1655-1
21. Askanas V, Kwan H, Alvarez RB et al (1987) De novo neuromuscular junction formation on human muscle fibres cultured in monolayer and

innervated by foetal rat spinal cord: ultrastructural and ultrastructural–cytochemical studies. J Neurocytol 16(4):523–537

22. Lambernd S, Taube A, Schober A et al (2012) Contractile activity of human skeletal muscle cells prevents insulin resistance by inhibiting pro-inflammatory signalling pathways. Diabetologia 55(4):1128–1139. https://doi.org/10.1007/s00125-012-2454-z
23. Nikolic N, Aas V (2019) Electrical pulse stimulation of primary human skeletal muscle cells. Methods Mol Biol 1889:17–24. https://doi.org/10.1007/978-1-4939-8897-6_2
24. Henry RR, Abrams L, Nikoulina S et al (1995) Insulin action and glucose metabolism in nondiabetic control and NIDDM subjects. Comparison using human skeletal muscle cell cultures. Diabetes 44(8):936–946

Check for updates

Chapter 12

Extracellular Vesicle Isolation and Purification from Skeletal Muscle Cells

Dongnan Yan, Emily Haughton, Ruhani Makkar, Matthew J. A. Wood and Dhanu Gupta

Abstract

Extracellular vesicles (EVs) have emerged as critical mediators of intercellular communication and hold considerable promise as biomarkers and therapeutic vectors. In skeletal muscle, EVs transport proteins, RNA, and other bioactive molecules, such as myokines, making them valuable indicators of muscle physiology and pathology. This method paper presents a detailed protocol for the isolation, purification, and characterization of EVs derived from human myoblasts. Our optimized workflow ensures high yield and sample quality, enabling comprehensive downstream molecular analysis. Additionally, we describe advanced techniques for assessing EV physical characteristics, concentration, and cargo content, providing an essential toolkit for investigating skeletal muscle EV biology and potential role in disease diagnostics. This protocol facilitates the efficient exploration of skeletal muscle-derived EVs and their implications in muscle function and pathology.

Keywords Skeletal muscle, Extracellular vesicles, EV isolation, EV purification, EV characterization, Ultracentrifugation

1 Introduction

EVs have gained interest in the field of medicine over recent years due to their potential as biomarkers and vectors for delivering therapeutic cargo. These nanosized, membrane-bound vesicles are produced endogenously by nearly all types of cells as part of the Endosomal Sorting Complex Required for Transport (ESCRT) pathway [1]. Specifically, they are released from either the fusion of multivesicular bodies (MVB) with the plasma membrane, or through the outward budding of the plasma membrane [2] (Fig. 1). EVs can be categorized by size into exosomes, microvesicles, and apoptotic bodies, each having unique biogenesis pathways and functions.

The authors Dongnan Yan and Emily Haughton contributed equally.

Joe Quadrilatero (ed.), *Myogenesis and Skeletal Muscle Biology*, Methods in Molecular Biology, vol. 3057,
https://doi.org/10.1007/978-1-0716-5396-8_12,

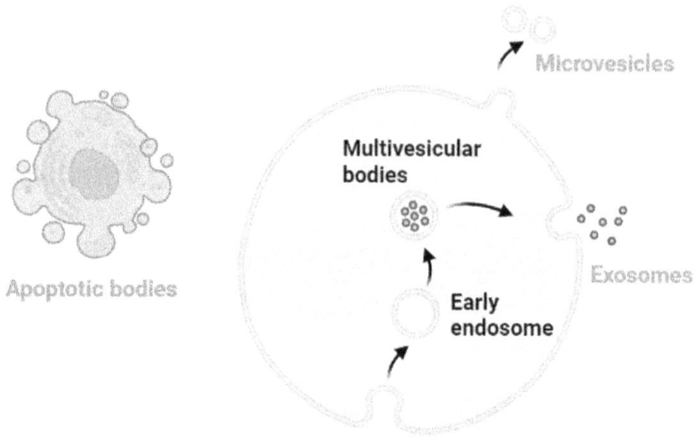

Fig. 1 Biogenesis of EVs. EVs are categorized into Exosomes (30–150 nm), which form inside multivesicular bodies (MVBs) and are released when MVBs fuse with the plasma membrane, microvesicles (100–1000 nm), which shed directly from the plasma membrane through outward budding, and apoptotic bodies (500–2000 nm), which are released during apoptosis

A key function of EVs is mediating intercellular communication by transporting bioactive cargo, such as RNA, proteins, and membrane receptors. This capacity makes them ideal candidates for both disease biomarkers and use in therapeutics [3]. In this method, we focus on EVs secreted from human skeletal muscle, which accounts for approximately 40% of human muscle mass. Like other cell types, human myoblasts also actively produce EVs [4, 5]. In response to contraction, myokines can be secreted through the classical signaling pathway and packaged within EVs [6]. Notably, approximately 400 proteins have been identified in EVs from skeletal muscle, including many well-established myokines [7]. Both undifferentiated and differentiated myoblasts have been shown to secrete EVs that can modulate recipient cell function, including influencing differentiation state [8].

EVs derived from skeletal muscle are also relevant in disease. They carry specific cargos, both influencing pathogenesis of disease [9] and giving indications of the pathological state of the skeletal tissue [10]. Analysis can also be used to gain insight into the molecular profile of these EVs, holding potential for skeletal disease mapping and biomarker discovery. This chapter will detail a comprehensive pipeline for the production, purification, and characterization of EVs from human skeletal myoblast cells (Fig. 2). This protocol has been optimized to maximize efficiency, minimize vesicle loss, and ensure high sample quality for downstream analysis.

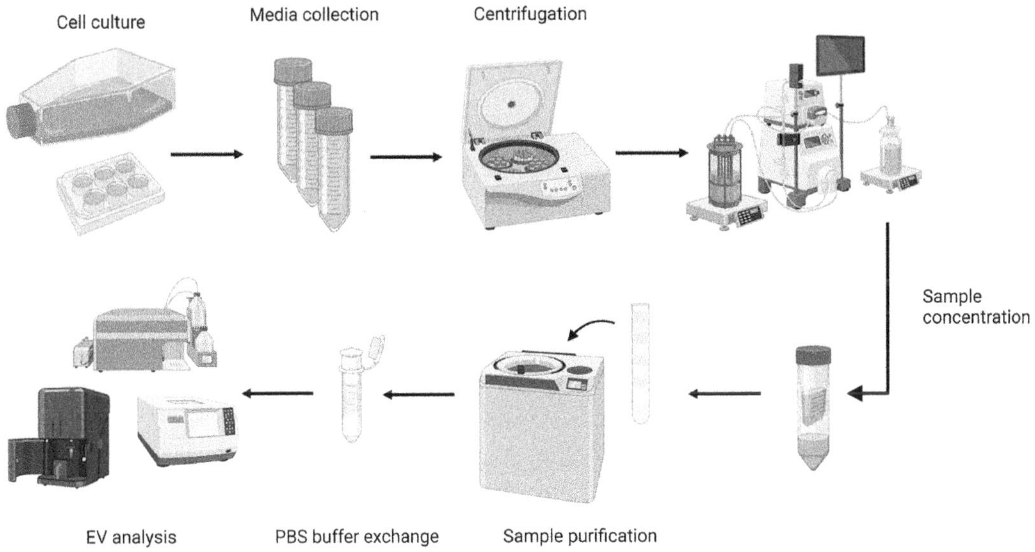

Fig. 2 A step-by-step process for preparing and analyzing extracellular vesicles (EVs) from cell culture is shown. It begins with growing cells in culture flasks or plates, followed by collecting the culture media in tubes. The media is then centrifuged to separate components. After centrifugation, the sample undergoes concentration using a filtration device. The concentrated sample is purified by another centrifugation step with a purification column, followed by exchanging the buffer to phosphate-buffered saline (PBS) in a small tube. Finally, the purified sample is analyzed using specialized instruments for EV characterization. Arrows indicate the flow from one step to the next, highlighting the preparation and analysis sequence needed to isolate and study EVs.

2 Materials

2.1 Reagents

1. Human immortalized skeletal muscle cells (e.g., MRC CNMD Biobank London, L954/1284 M-I).
2. Complete growth media: Skeletal Muscle Cell Growth Medium (low serum 5%V/V) + supplements listed below [3–8].
3. Fetal Calf Serum: 0.05 mL/mL.
4. Fetuin (bovine): 50 µg/mL.
5. Epidermal Growth Factor (recombinant human): 10 ng/mL.
6. Basic Fibroblast Growth Factor (recombinant human):1 ng/mL.
7. Insulin (recombinant human): 10 µg/mL.
8. Dexamethasone: 0.4 µg/mL.
9. PBS pH 7.4 (−)$CaCl_2$ (−)$MgCl_2$.
10. Dissociation reagent: gibco TrypLE express enzyme (−)phenol red (+)EDTA.
11. Differentiation media: Skeletal muscle differentiation media + supplements [Insulin (recombinant human): 10 µg/mL].

12. Reduced serum media: Gibco Opti-MEM (1×) (1% antibiotic/antimycotic solution).
13. Antibiotic/antimycotic solution: 10,000 units penicillin, 10 mg streptomycin, 25 μg amphotericin B/mL.
14. 0.22 μm syringe filters and syringes.
15. MiliQ water.
16. NanoFCM™ Quality Control Nanospheres (QC beads) and NanoFCM™ Silica Nanospheres Cocktail (sizing beads).
17. ONI EV profiler/ONI EV profiler 2 kits.
18. RNaseZap.
19. TRIzol™ LS Reagent.
20. Chloroform.
21. Isopropanol.
22. Glycogen.
23. Glycerol.
24. 70% Ethanol.
25. RNase-free water.
26. TURBO DNA-free Kit.
27. High-Capacity cDNA Reverse Transcription Kit.
28. DdPCR supermix and consumables.
29. RIPA buffer.
30. 50× proteinase inhibitor.
31. Pierce™ BCA Protein Assay Kit.
32. NuPAGE LDS Sample Buffer (4×).
33. NuPAGE Sample Reducing Agent (10×).
34. NuPAGE Bis–Tris Protein Gels, 4%–12%.
35. NuPAGE MOPS SDS Running Buffer.
36. Trans-Blot Turbo Transfer Packs.
37. SuperBlock Blocking Buffer.
38. Restore Western Blot Stripping Buffer.
39. TBST made by TBS 10× Stock Solution and Tween 20.
40. OptiPrep Density Gradient Medium (Iodixanol solution).
41. 5× PBS-MK: 131.5 mg of $MgCl_2$ and 74.55 mg of KCl in 100 mL of 5× PBS.

2.2 Equipment

1. Beckman Coulter Optima XPN-80 Ultracentrifuge equipped with SW 41 Ti Swinging-Bucket Rotor.
2. Beckman Coulter 13.2 mL Open-Top Thinwall Ultra-Clear Tube, 14 × 89 mm.

3. Tangential Flow filtration system (e.g., Repligen KrosFlo® KR2i).
4. Amicron Ultra spin filter columns.
5. Ibidi μ-Slide 8 Well high chamber slide.
6. NanoSight NS300.
7. NanoFCM™ Nanoflow Cytometer.
8. Vortex mixer.
9. Super-resolution microscope (ONI).
10. Centrifuge.
11. Heating block.
12. NanoDrop.
13. PX1 PCR Plate Sealer.
14. Thermal Cycler.
15. QX200 Droplet Digital PCR System.
16. Western Blot imaging system.
17. Trans-Blot Turbo Transfer System.
18. Midikros 65cm 300k mPES 0.5mm fll x fll 1/pk sterile.

3 Methods

3.1 Cell Culture

1. Grow human skeletal myoblasts in complete growth media in a T175 flask until 80% confluent. Incubate cells at 37°C, 85% humidity and 5% CO_2.
2. Before subculturing, pre-warm media to 37°C.
3. Remove media. Wash three times with PBS (*see* **Note 1**).
4. Dissociate cells by adding 2 mL of dissociation reagent and incubating at 37°C for 5 min.
5. Gently agitate flask to lift cells before adding 8 mL of complete growth media to neutralize dissociation reagent.
6. Collect cells and add to a 50 mL Falcon tube (*see* **Note 2**).
7. Centrifuge at 500 × *g* for 5 min to pellet cells.
8. Remove supernatant and resuspend pellet in 10 mL of fresh complete growth media.
9. If wanting to differentiate, seed into 6-well plates at 3×10^5 cells per well (2 mL of complete growth media per well).
10. Once cells are 80% confluent, change media to differentiation media (2 mL per well).
11. Every 2 days, remove media, wash with PBS, and add 2 mL of fresh differentiation media per well.

12. The differentiation process can take a couple of days to a few weeks depending on the condition of the cells. (*see* **Note 3**).
13. Once differentiated, change media to reduced serum media and leave for 48 h. (*see* **Note 10, 20, 21 & 26**).
14. Collect the media into 50 mL Falcon tubes at 300 × *g* for 5 min, followed by 4000 × *g* for 10 min to pellet any debris or residual cells. Change tubes between the two spins.

3.2 Filtration

1. The initial concentration is done using the Tangential Flow Filtration system. The lab uses KrosFlo® KR2i TFF System; however, the steps described further are similar in sequence for different filtration systems.
2. Ensure the pore size of the column does not exceed 350 kDa, as this would cause a loss of sample into the waste. (*see* **Note 12**).
3. The machine needs to be connected to the columns with the filtrate and retentate tubes plugged as described in the user's manual from the manufacturer.
4. Select the appropriate program according to the instructions provided by the manufacturer.
5. Run two column volumes of PBS to rinse the column. (*see* **Note 11**).
6. Run the sample to concentrate it down to ~15–20 mL. If the sample concentrate becomes viscous, add PBS in 1:1 ratio to avoid chances of precipitation of the sample.
7. Between running samples, clean the column by running two column volumes of MilliQ distilled water, two column volumes of 0.5 M NaOH, followed by two more column volumes of MilliQ.
8. Before loading the next sample, prime the column by running two column volumes of PBS.
9. Once all samples have been run, repeat column clean; however, finish with two column volumes of 20% glycerol rather than PBS. Once the column is stored in 20% glycerol, secure the valves and disconnect the tubing. The column can now be kept at 4°C.
10. To concentrate the media down further, the use of spin filters with 100 kDa molecular weight cut off (MWCO) is recommended.
11. Prime the spin filters by running one column volume of PBS before adding your sample and proceed to centrifuge at 2000 × *g* until the final volume of 0.5 mL is achieved.

12. As the particles are likely to clog the filter pores, agitating the sample by pipetting up and down occasionally is highly recommended. (*see* **Note 13 & 14**).
13. The sample is now ready for iodixanol gradient ultracentrifugation.

3.3 Density-Gradient Fractionation

1. Pre-prepare the iodixanol fractions as described in Table 1.
2. Layer 1 mL of each iodixanol solution sequentially in a 13.2 mL Open-Top Thin wall Ultra-Clear Tube, following the order specified in Fig. 3. After adding the final iodixanol fraction, gently add the concentrated sample on top of the gradient. (*see* **Note 15**).

Table 1

Component composition for the preparation of iodixanol gradients. Top up each fraction with iondixanol to 50 ml

Iodixanol	5 M NaCl	5× PBS-MK	H_2O	Phenol red
6%	10 mL	10 mL	25 mL	–
	–	10 mL	30 mL	100 μL
14%	–	10 mL	28.3 mL	–
16%	–	10 mL	26.67 mL	100 μL
18%	–	10 mL	25 mL	–
25%	–	10 mL	19.2 mL	100 μL
40%	–	10 mL	6.7 mL	–
60%	–	10 mL	–	100 μL

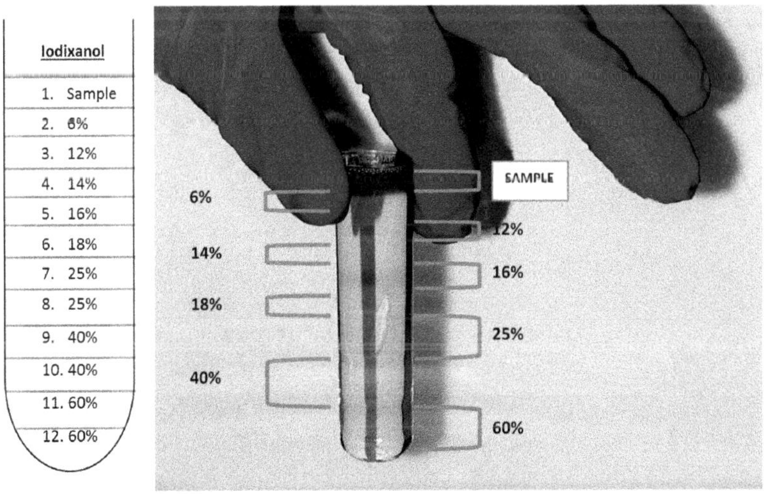

Fig. 3 Visual representation of iodixanol density gradient fractions. Layered iodixanol gradient used for EV isolation by ultracentrifugation. Discrete 1 mL layers of iodixanol ranging from 6% to 60% are sequentially loaded into an ultracentrifuge tube as shown. The EV-containing sample is placed above the 6% layer

3. Run the samples on the Ultracentrifuge for 16 h at 110,000 × *g* in the SW 41 Ti Swinging-Bucket Rotor.
4. After the run has ended, collect separate 1 mL fractions layer by layer starting from the top of the tube, placing the samples immediately on ice. (*see* **Note 16 &18**).
5. Based on previous downstream analysis of individual fractions done in the lab, gradients 4, 5, and 6 have been identified as EV-enriched and can, therefore, be directly selected for further EV characterization. (*see* **Note 19 & 24**).

3.4 EV Analysis

3.4.1 Physical Characteristics of EVs

A) **NTA**

1. Dilute EV sample in filtered PBS to achieve a concentration suitable for NTA, typically resulting in 20–100 particles per frame for optimal analysis.
2. Perform measurements using a NanoSight NS300 equipped with a 488 nm laser.
3. Calibrate the instrument using known size standards (e.g., 100 nm polystyrene beads) before measurements to ensure accuracy.
4. Introduce samples into the flow cell using a syringe pump set at a constant flow rate to ensure even distribution of particles during analysis.
5. For each sample, capture three 60-s video recordings at room temperature, with each video containing between 30 and 60 frames per second.
6. Analyse video data using the instrument's software (e.g., NTA 3.2), which tracks the Brownian motion of individual vesicles to calculate their hydrodynamic diameter. The software generates particle size distribution profiles, providing both size and concentration data for the EV sample.

B) **NanoFCM**

Nano-Flow Cytometry (NanoFCM) is a highly sensitive technique used to analyze EVs at the single-particle level. It allows for accurate size determination and concentration, providing critical insights into EV heterogeneity. This method is based on nanoparticle fluorescence and light scattering, enabling both quantitative and qualitative analysis.

1. After the isolation and purification of EVs, dilute EV samples in 0.22 μm membrane-filtered PBS to a concentration of approximately 1×10^9 particles/mL.
2. Perform calibration using standard nanoparticles of known sizes and fluorescence (e.g., silica nanoparticles) to ensure

accurate size and concentration measurements, and to standardize the fluorescence intensity across channels.

3. Run blank PBS samples to check for background noise or contamination.
4. Set the detection threshold to discriminate EVs from background noise and smaller particles.
5. Load the prepared EV sample into the NanoFCM system and run according to the instructions provided by the manufacturer (*see* **Note 17**).
6. The size distribution and concentration of the sample will be reported by the software connected to NanoFCM once calibrated by appropriate quality control and sizing beads.
7. Ensure reproducibility by comparing EV samples from biological replicates.

C) **Super-Resolution Microscopy (ONI)**

Super-resolution microscopy, particularly using the ONI (Oxford Nanoimaging) platform, enables visualization of EVs with nanometre-scale precision, surpassing the diffraction limit of conventional microscopy [11]. This technique allows for detailed structural and spatial analysis of EVs, including membrane composition and the localization of specific surface or cargo proteins. By combining fluorescence and high-resolution imaging, ONI provides critical insights into EV heterogeneity and molecular content.

1. After isolation and purification of EVs, dilute the EV sample to a concentration suitable for single-particle imaging (typically ~1×10^9 particles/mL). (*see* **Note 23 & 25**).
2. Prepare the samples in an 8-well chamber or a slide provided by ONI EV Profiler kit according to the instructions suggested by ONI.
3. Turn on the ONI super-resolution microscope and allow sufficient time for the laser warm-up and stabilize.
4. Choose appropriate laser excitation wavelengths depending on the fluorescent probes used (e.g., 488 nm for green dyes, 561 nm for red dyes).
5. Set up the detection channels for simultaneous multi-color imaging if analyzing multiple markers.
6. Focus the immobilized EVs and adjust the z-plane to bring the sample into optimal focus.
7. Start image acquisition by activating the lasers and capturing multiple frames (~10,000–50,000 frames) for localization microscopy.

8. For each frame, track individual fluorescent molecules over time to build a super-resolved image of the EVs.
9. Ensure sufficient photon counts are collected per molecule to achieve the desired localization precision.
10. Use ONI's software CODI to reconstruct the super-resolution image by fitting Gaussian profiles to the fluorescent spots in each frame.
11. Analyse the spatial distribution of EVs and quantify vesicle size with nanometre resolution.
12. Include control samples, such as fluorescently labelled beads or unstained EVs, to verify system calibration and imaging accuracy.
13. Ensure image acquisition is performed under conditions that prevent photobleaching or photo-damage to the sample.
14. Repeat imaging across multiple fields of view and replicate samples to ensure reproducibility.

3.4.2 Characterization of Loaded RNA

A) **RNA Extraction from EVs**

1. Use RNaseZap to sterilize the working surface.
2. After measuring EV concentration, add at least 5×10^{10} isolated EVs in a 1.5 mL Eppendorf tube, and add 0.22 μm filtered PBS to top up to 250 μL.
3. Add 750 μL of TRIzol LS Reagent. Pipette up and down to mix well. Incubate 5 min in room temperature.
4. Add 200 μL of Chloroform, cap the tube. Vortex vigorously for 15 s and incubate 10 min in RT (*see* **Note 4**).
5. Centrifuge the sample at 14,000 × *g* for 15 min at 4°C.
6. After centrifugation, the mixture separates into three phases. Transfer the aqueous phase carefully to a new tube.
7. Add 500 μL isopropanol and 1 μL of glycogen. Vortex well. Incubate for 10 min at room temperature (*see* **Note 5**).
8. Centrifuge 14,000 × *g* for 15 min at 4°C. A gel-like pellet will form at the bottom of the tube.
9. Discard the supernatant and resuspend the pellet in 1 mL pre-chilled 80% ethanol.
10. Centrifuge 7500 × *g* at 4°C for 5 min.
11. Air dry the pellet
12. Resuspend the pellet in RNase-free water. Use NanoDrop to measure the concentration of the RNA (*see* **Note 6**).

B) **Reverse Transcription and Digital droplet PCR (ddPCR)**

1. Reverse transcription is done following the protocol of High-Capacity cDNA Reverse Transcription Kit. The concentration of cDNA is measured by NanoDrop spectrophotometer.
2. The ddPCR is performed according to the manufacturer's instructions using ddPCR Supermix for Probes/Evagreen. The probes/primers are designed based on the sequence to be detected. The copy number of the target sequence will be directly quantified using QX Manager Software.

3.4.3 Characterization of Proteins on/Within EVs

A) **Characterization of Essential EV Markers**

The Minimal Information for Studies of Extracellular Vesicles (MISEV) guidelines, developed by the International Society for Extracellular Vesicles (ISEV), set recommendations for identifying and characterizing EVs [12]. These guidelines emphasize the importance of using a combination of markers to ensure accurate identification of EVs and to avoid contamination from non-vesicular components. The MISEV guidelines recommend using at least three positive markers (with one cytosolic protein) and one negative marker to ensure the vesicles being studied are indeed EVs. Following are some recommendations for marker selection. (*see* **Note 22**).

i. Transmembrane or GPI-anchored proteins: Proteins-like tetraspanins (e.g., CD9, CD63, CD81) are often associated with EVs and are located on their surface.
ii. Cytosolic proteins: Proteins-like TSG101 or Alix, which are involved in EV biogenesis, are common in the lumen of EVs.
iii. Negative markers: Endoplasmic reticulum proteins (e.g., calnexin), Golgi apparatus proteins, etc.

Western blot can be used to identify the selected EV markers:

1. EV producer cells are lysed with RIPA buffer supplemented with a proteinase inhibitor. Incubate the lysate on ice for 30 min and mix gently every 5–10 min.
2. Centrifuge the lysate at 10,000 × *g* for 15 min at 4°C. Carefully transfer the supernatant to a new tube.
3. The concentration of the extracted protein can be quantified using a Pierce BCA Protein Assay Kit.
4. Mix cell-extracted protein/EV sample with NuPAGE LDS Sample Buffer (4×) and NuPAGE Sample Reducing Agent (10×) and incubate at 95°C for 10 min. See Table 2.
5. Load sample to a NuPAGE Bis–Tris Protein Gel and run western blot following the instructions provided by the manufacturer (*see* **Note 7**).

Table 2
An example of a component recipe for a 16 μL reaction

Cell protein sample	
Components	**Volume**
NuPAGE LDS Sample Buffer (4×)	4 μL
NuPAGE Sample Reducing Agent (10×)	1.6 μL
Protein (5–10 μg)	*X* μL
Lysis buffer (RIPA buffer with proteinase inhibitor)	(10.4 − *X*) μL
Total	*16μL*
EV sample	
Components	**Volume**
NuPAGE LDS Sample Buffer (4×)	4 μL
NuPAGE Sample Reducing Agent (10×)	1.6 μL
Purified EV (4×10^8–4×10^9 particles)	*X* μL
PBS	(10.4 − *X*) μL
Total	*16μL*

B) **Antibody labeling of membrane protein**

1. Fluorescent antibody of the target protein can be pre-diluted with 0.22 μm-filtrated PBS at 1:10, 1:100, 1:1000, and 1:10,000.
2. Dilute the EV sample to 1×10^9 particle/mL using 0.22 μ m-filtrated PBS, transfer 100 μL EV sample to each 0.6 mL Eppendorf tube.
3. Add pre-diluted antibody to each tube. Wrap the tube with foil paper and fix them to a rotator mixer for overnight labeling.
4. NanoFCM can be used to characterize the percentage population of fluorescent/double-fluorescent particles (*see* **Note 8**).
5. Super-resolution microscope ONI can also be used to colocalize up to three fluorescent proteins in the particle (*see* **Note 9**).

C) **Characterization of soluble proteins**
Western blotting can also be applied to relatively quantify proteins other than the MISEV markers (e.g., protein cargos, other membrane proteins [12]). ImageJ can be used to quantify the intensity of the band. If the target protein is fluorescent, it

can also be analyzed by NanoFCM or ONI. If not fluorescent, EVs can be fixed and permeabilized to allow the fluorescent antibody to be taken up and label the target protein following the instructions of ONI EV profiler 2 kit.

4 Notes

1. Always aspirate PBS from the same corner when washing cells to avoid disturbing the monolayer.
2. Tilt the flask and rinse cells down the surface to ensure efficient recovery during detachment.
3. Differentiated myoblasts exhibit elongated, fused morphology—monitor closely for consistency.
4. TRIzol LS Reagent and Chloroform must be handled in fume hood.
5. For better yield, it is recommended to incubate the mixture at −20°C for 2 h.
6. The TRIzol-extracted RNA might have DNA contamination. It is highly recommended that the RNA is further purified with a TURBO DNA-free™ Kit following the instructions by the manufacturer.
7. It is recommended to make a mixture of the primary antibodies for the MISEV markers.
8. NanoFCM only has two lasers and, therefore, can only detect up to two fluorescent signals at the same time.
9. EV membrane markers can also be visualized using super-resolution microscope ONI, following the instructions using the ONI EV profiler kit.
10. Use reduced serum media (e.g., Opti-MEM) for 24–48 h to minimize bovine EV contamination.
11. Ensure TFF columns are flushed with PBS before use to remove glycerol storage buffer.
12. Avoid pore sizes >350 kDa in TFF columns to prevent EV loss during concentration.
13. Gently agitate spin filters during ultrafiltration to avoid membrane clogging by vesicles.
14. Limit final EV concentrate to ~0.5 mL to prevent aggregation or protein precipitation.
15. Pipette iodixanol gradients carefully to maintain sharp interfaces between layers.
16. Always place EV-enriched fractions on ice immediately after ultracentrifugation.

17. Use biological replicates and include PBS-only controls to ensure EV purity and reproducibility.
18. Perform all EV-related procedures using RNase-free and low-retention plasticware.
19. Avoid repeated freeze–thaw cycles of EV samples to preserve vesicle integrity and function.
20. Pre-warm all media and buffers to 37°C before adding to cells to maintain physiological conditions.
21. For loading EVs with cargo, perform transfection at ~80% confluency and switch to Opti-MEM after 4 h.
22. Validate fraction purity using MISEV-recommended markers before downstream analysis.
23. Include blank controls (PBS processed through entire workflow) to monitor background signals.
24. Store purified EVs at −80°C in low-protein binding tubes and thaw gently on ice before use.
25. Use phenol red-free media for EV imaging studies to reduce background fluorescence.
26. Document batch numbers of serum and media supplements, as EV content can vary significantly.

References

1. Radler J, Gupta D, Zickler A, Andaloussi SE (2023) Exploiting the biogenesis of extracellular vesicles for bioengineering and therapeutic cargo loading. Mol Ther 31(5):1231–1250
2. Abels ER, Breakefield XO (2016) Introduction to extracellular vesicles: biogenesis, RNA cargo selection, content, release, and uptake. Cell Mol Neurobiol 36(3):301–312
3. Kalluri R, LeBleu VS (2020) The biology, function, and biomedical applications of exosomes. Science 367(6478), eaau6977.
4. Zanotti S, Gibertini S, Blasevich F, Bragato C, Ruggieri A et al (2018) Exosomes and exosomal miRNAs from muscle-derived fibroblasts promote skeletal muscle fibrosis. Matrix Biol 74:77–100
5. Aoi W, Tanimura Y (2021) Roles of skeletal muscle-derived exosomes in organ metabolic and immunological communication. Front Endocrinol (Lausanne) 12:697204
6. Trovato E, Di Felice V, Barone R (2019) Extracellular vesicles: delivery vehicles of myokines. Front Physiol 10:522
7. Bydak B, Pierdona TM, Seif S, Sidhom K, Obi PO, et al (2022) Characterizing extracellular vesicles and particles derived from skeletal muscle myoblasts and myotubes and the effect of acute contractile activity. Membranes (Basel) 12(5):464.
8. Choi JS, Yoon HI, Lee KS, Choi YC, Yang SH et al (2016) Exosomes from differentiating human skeletal muscle cells trigger myogenesis of stem cells and provide biochemical cues for skeletal muscle regeneration. J Control Release 222:107–115
9. Miao C, Zhang W, Feng L, Gu X, Shen Q et al (2021) Cancer-derived exosome miRNAs induce skeletal muscle wasting by Bcl-2-mediated apoptosis in colon cancer cachexia. Mol Ther Nucleic Acids 24:923–938
10. Koutsoulidou A, Photiades M, Kyriakides TC, Georgiou K, Prokopi M et al (2017) Identification of exosomal muscle-specific miRNAs in serum of myotonic dystrophy patients relating to muscle disease progress. Hum Mol Genet 26(17):3285–3302
11. Xu S, Zhang Z, Melvin BC, Basu Ray N, Ikezu S, Ikezu T (2024) Comparison of nanoimaging and nanoflow based detection of extracellular vesicles at a single particle resolution. J Extracell Biol 3(10):e70016
12. Welsh JA, Goberdhan DCI, O'Driscoll L, Buzas EI, Blenkiron C et al (2024) Minimal information for studies of extracellular vesicles (MISEV2023): from basic to advanced approaches. J Extracell Vesicles 13(2):e12404

Check for updates

Chapter 13

Flow Cytometric Methods for ROS Quantification in Skeletal Muscle Myotubes

Andy Nolan, Paige Cole and Darren W. Sexton

Abstract

Flow cytometry is a high-throughput technique which can be used to analyze and quantify characteristics of individual cells. Flow cytometry hydrodynamically focuses cells past a series of size and fluorescent detectors to characterise cellular morphology and phenotype. This process of hydrodynamic focusing has previously been shown to be incompatible with myotubes due to their larger size relative to other cell types. Here, we describe the culture and differentiation of myotubes and how myotubes can be analyzed via flow cytometry to quantify changes in cellular and mitochondrial-specific ROS production.

Keywords Flow cytometry, Skeletal muscle, C2C12 myotubes, Reactive oxygen species, Mitochondria

1 Introduction

Skeletal muscle contributes approximately 40%–50% of healthy young adult mass [1]. There exist ~650 different skeletal muscles within the human body, and they are essential for a variety of different roles, including producing movement, sustaining posture, maintenance of body temperature, storage of nutrients, and provision of energy [2]. Mechanically, skeletal muscle's primary function is the conversion of chemical energy into mechanical energy via the breakdown of the high-energy bond within the molecule adenosine triphosphate (ATP). Therefore, skeletal muscle is highly energetic, accounting for approximately 30% of whole-body energy expenditure at rest, which can increase 100-fold following the onset of exercise [1, 3, 4].

These bioenergetic demands can be met following the breakdown of substrates, such as glucose or fatty acids, in both anaerobic and aerobic processes. The anaerobic process of glycolysis involves a sequential set of reactions, which convert glucose into pyruvate, generating four ATP molecules in the process. If oxygen is present, then pyruvate can be further metabolized within the mitochondria in a process known as the tricarboxylic acid cycle (TCA cycle),

Joe Quadrilatero (ed.), *Myogenesis and Skeletal Muscle Biology*, Methods in Molecular Biology, vol. 3057,
https://doi.org/10.1007/978-1-0716-5396-8_13,

producing two more ATP, alongside reducing NAD^+ and FAD to NADH and $FADH_2$, respectively. Fatty acid ATP production also exclusively occurs within the mitochondria in a process known as β-oxidation, which cyclically cleaves off two carbons of the fatty acid chain, producing acetyl-CoA, NADH and $FADH_2$ per cycle. These reducing equivalents can then be oxidized by large mitochondrial protein complexes within the electron transport chain, which traffic the electrons produced by the oxidation of NADH and $FADH_2$ via a series of electron carriers to the terminal electron acceptor water, while also generating the electrochemical gradient required to produce ATP. This process is extremely efficient, and the oxidative metabolism of glucose produces substantially more ATP than glycolysis, with two produced per TCA cycle and approximately 34 from oxidative phosphorylation. However, while efficient, this process is still bioenergetically imperfect and consequently electrons can leak (~0.1%–2%) to be subsequently captured by molecular water [5–7]. This produces what is known as reactive oxygen species (ROS), which are highly reactive molecules that contain an unpaired electron in their outer shell. Unsurprisingly, the mitochondria are, therefore, a major source of ROS within the cell, and the primary sources of ROS generation within the mitochondria are the ubiquinone-binding sites in Complex I (IQ) and Complex III (site IIIQo) [8]. However, there are also non-mitochondrial sources of ROS, such as nicotinamide adenine dinucleotide phosphate oxidase (NADPH oxidase), which has been shown in skeletal muscle to be the primary source of ROS during moderate-intensity exercise [9].

There exist a variety of ROS which differ in their reactivity. When an unpaired electron reduces molecular water, it forms a radical known as superoxide ($O^{2\bullet-}$). $O^{2\bullet-}$ is highly reactive with a brief half-life (10^{-6} -10^{-3} s) and is, therefore, converted to the more stable hydrogen peroxide (H_2O_2) by the enzyme superoxide dismutase [10]. H_2O_2 is more stable, has a longer half-life (few seconds), and is capable of crossing membranes via aquaporins [11]. Consequently, H_2O_2 can act as a secondary messenger by oxidizing thiol groups (–SH) of cysteine residues on proteins, leading to post-translational modifications. However, during pathological conditions, there can be an increase in ferrous iron (Fe^{2+}), which can interact with H_2O_2 to form extremely reactive hydroxyl radicals (•OH) in what is known as the Fenton reaction. •OH can then target the carbon–carbon double bonds, especially present within polyunsaturated fatty acids, causing lipid peroxidation, which can subsequently cause cell death via ferroptosis [12]. To combat this, cells possess innate antioxidant defence systems, such as glutathione peroxidase (GPx), which can detoxify H_2O_2 through the donation of electrons from two reduced glutathione molecules (GSH), forming H_2O and oxidized glutathione (GSSG). Therefore, while ROS play a crucial role in a variety of cellular processes, including signalling, pathogen defence, and ferroptosis, aberrant ROS signalling can overwhelm the cells' antioxidant capabilities and subsequently lead to tissue damage known as oxidative stress [13]. Given that skeletal muscle has a large bioenergetic demand which can lead to substantial changes within the cellular redox balance, the quantification of cellular ROS concentrations can help to better interrogate these changes.

This chapter will outline how flow cytometry can be used to quantify live cellular changes in both global cellular and mitochondrial-specific ROS production within cultured skeletal muscle.

2 Materials

2.1 Cell Expansion and Storage

1. Sterile, vented T75 Falcon, Corning, or Nunc.
2. Growth Medium (GM): Dulbecco's Modified Eagle's Medium (DMEM) with 4500 mg/L glucose and 2.5 mM L-glutamine 20% fetal bovine serum (FBS), 1% penicillin–streptomycin solution (PS) (50 units penicillin/50 μg streptomycin) stored at 4°C and prewarmed to 37°C prior to use.
3. Differentiation Medium (DM): DMEM with 4500 mg/L glucose and 2.5 mM L-glutamine, 2% Horse Serum (HS), 1% (PS) stored at 4°C and prewarmed to 37°C prior to use.
4. 0.01 M phosphate-buffered saline (1× PBS) stored and utilized at room temperature.
5. 0.5% Trypsin-0.2% EDTA stored at 4°C and used at 37°C.
6. Neubauer hemocytometer.
7. Sterile tissue culturing hood appropriate for expansion and proliferation of immortalized cell lines, e.g., Kojair Biowizard Silverline class II hood, alongside an appropriate sterile incubator maintained at 37°C and 5% CO_2.

2.2 Flow Cytometry

1. BD Accuri C6 flow cytometer which has channels for measuring fluorescent intensity (FL-1, FL-2, FL-3, and FL-4).
2. Changes in total cellular ROS concentrations can be assessed using the general ROS indicator probe chloromethyl 2′, 7′ -dichlorodihydrofluorescein diacetate (CM-H2DCFDA) (Thermo Scientific Inc., MA, USA) at a final concentration of 1 μM.
3. Mitochondrial-specific ROS production can be assessed using the MitoSOX Red fluorescent reporter at a final concentration of 5 μM (Thermo Scientific Inc., MA, USA).

3 Methods

3.1 Myotube Population Expansion and Differentiation

3.1.1 Passaging Skeletal Muscle Cells

1. Source Murine C2C12 skeletal muscle myoblasts from the American Tissue Culture Collection (ATCC; Rockville, USA).
2. Passage cells to increase yield and store in liquid nitrogen (LN_2) until required for experimentation. Culture C2C12 myoblasts in T75 flasks containing 15 mL GM. The T75 flask should then be placed in the incubator (37°C, 5% CO_2) to proliferate until ~80% confluency is attained.
3. Once ~80% confluent, existing GM can be aspirated, and cells should be washed twice with 1× PBS.
4. Add 1 mL of trypsin EDTA to the T75 flask, ensuring full coverage, and incubate for 5 min (37°C, 5% CO_2) to dissociate cells that form the culture flask surface.
5. Following the detachment of cells, add 4 mL of GM to the flask to neutralize trypsin activity and centrifuge cells at 300 ×*g* for 5 min.
6. Remove the trypsin/GM solution and resuspend in GM for cell counting.

3.1.2 Cell Counting by Trypan Blue Exclusion

1. Cells can be manually counted using the trypan blue exclusion method, whereby 20 μL of cell suspension is mixed with 20 μL 0.4% trypan blue stain in a 1:1 ratio.
2. The solution should then be pipetted onto either end of the hemocytometer, flooding both chambers via capillary action. Cells in the four corner grids should be counted under a microscope at ×10 magnification. Viable cells will be recognized as small, round, and clearly visible, while non-viable cells are misshapen, slightly larger, with reductions in their membrane integrity, hence were trypan blue positive and excluded from cell counts.
3. Calculate the resultant means of the four grids, which represent average cell numbers per 0.1 mm^3 grid. Then multiply by 10^4 to extrapolate the number of cells in the 0.1 mm^3 to 1 cm^3 (equivalent to 1 mL of cell suspension). The total number of cells present within the cell suspension can then be calculated by simply multiplying the total volume of cell suspension (mL).

3.1.3 Differentiating C2C12 myoblasts

1. Monolayer C2C12 cultures should then be seeded on 6-well plastic tissue culture plates. C2C12 myoblasts should be seeded at 1×10^4 cells/cm^2 and cultured until ~80% confluency is achieved.

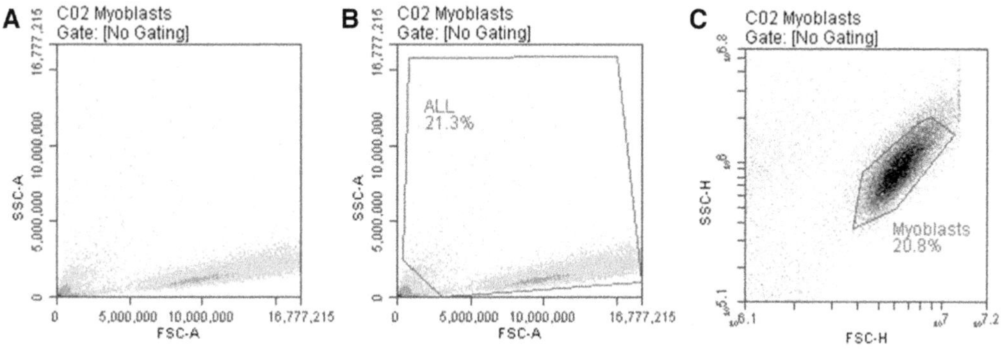

Fig. 1 FSC:SSC from a myoblast populations. (**A**) Myoblasts present a uniform FSC:SSC profile whereby the myotubes cluster together with low heterogeneity into a "cigar" shaped cluster. (**B**) Initial gating should attempt to include all available cell populations while removing the bottom left corner, which is comprised of small particles, debris, and instrumental noise. (**C**) Logarithmic gating of the "ALL" group will demonstrate a very distinct, tightly clustered group of cells for clear population identification

2. Aspirate remaining GM and wash monolayers twice with 1× PBS.
3. Add DM to monolayers to initiate the differentiation of myoblasts into multinucleated myotubes. The DM should be replaced every 48 h for a total of 8 days.
4. After 8 days in DM, clearly defined myotubes will be identifiable under phase contrast microscopy and the cultures are ready for experimentation.

3.2 Flow Cytometry Gating Strategy

1. The gating strategy will be described using the BD Accuri C6 Plus Analysis software for simpler visualization; however, gating and subsequent analysis will differ slightly depending on the instrument and can also be performed using external cytometry-specific analysis packages, such as FlowJo (BD, NJ, USA).
2. Differentiated myotubes, when analyzed on the flow cytometer, will display greater heterogeneity compared to more uniform cell types, such as myoblasts (*see* **Note 1**). Myoblasts, for instance, when analyzed using flow cytometry will display a clear single-cell population that is visually distinct (Fig. 1) (*see* **Note 2**).
3. Myotubes, however, present a scatterplot resembling (Fig. 2A). The FSC:SSC will typically present two distinct cell populations (although this can sometimes form into one large population depending on the sample). The morphologically smaller second population is thought to be apoptotic bodies, which are large enough to be detected using FSC:SSC (500 nm–2 μm). Moreover, the P2 population demonstrates greater Annexin V staining compared to P1, which is indicative of apoptosis (*see* **Note 3**) [14].

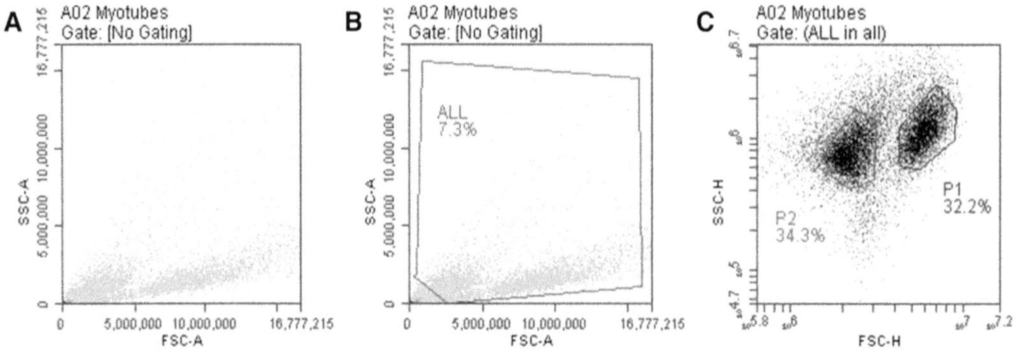

Fig. 2 FSC:SSC of myotube populations. (**A**) Myotubes display greater heterogeneity when compared to more uniform cell types such as myoblasts. Consequently, the cells typically cluster into two unique cell populations but can form one large, clustered group of cells. (**B**) Initial gating should attempt to gate around all the cells while excluding the bottom left corner, which is comprised of small particles, debris, and instrumental noise. (**C**) The "ALL" population should then be converted to logarithmic gating, which helps better visualize the two cell subpopulations. The small population (P2) has previously been shown to display high Annexin V staining and is possibly apoptotic bodies. Therefore, gating on the P1 population only is recommended

4. To better enhance the differences between the two populations, gating around the bottom left corner of the scatterplot will help to remove debris and instrumental noise (Fig. 2B), and a further plot on the logarithmic scale should be performed (Fig. 2C).
5. Logarithmic gating helps to further stratify the two cell populations with a distinct larger myotube P1 population and a much smaller P2 population.
6. This somewhat paradoxical FSC:SSC profile can be attributed to the tendency of myotubes to "roll up" during cytometric analysis and subsequently present as smaller during size detection (*see* **Notes 4 and 5**).

3.2.1 General ROS Staining

1. The CM-H2DCFDA probe is a cell-permeable dye which can be oxidized to produce a ROS-dependent fluorescent product at an excitation/emission maxima of 492–495/517–527 nm (*see* **Note 6**).
2. To quantify cellular ROS concentrations, remove the cell medium and wash cells twice in PBS before the adding of trypsin.
3. Following 5 min incubation at 37°C and visual confirmation of cell detachment, neutralize the trypsin via the addition of DM and spin cells at 300 rcf for 5 min.
4. The remaining media/trypsin can be removed with the resultant cell pellet resuspended in 1 μM CM-H2DCFDA solution (*see* **Note 7**).

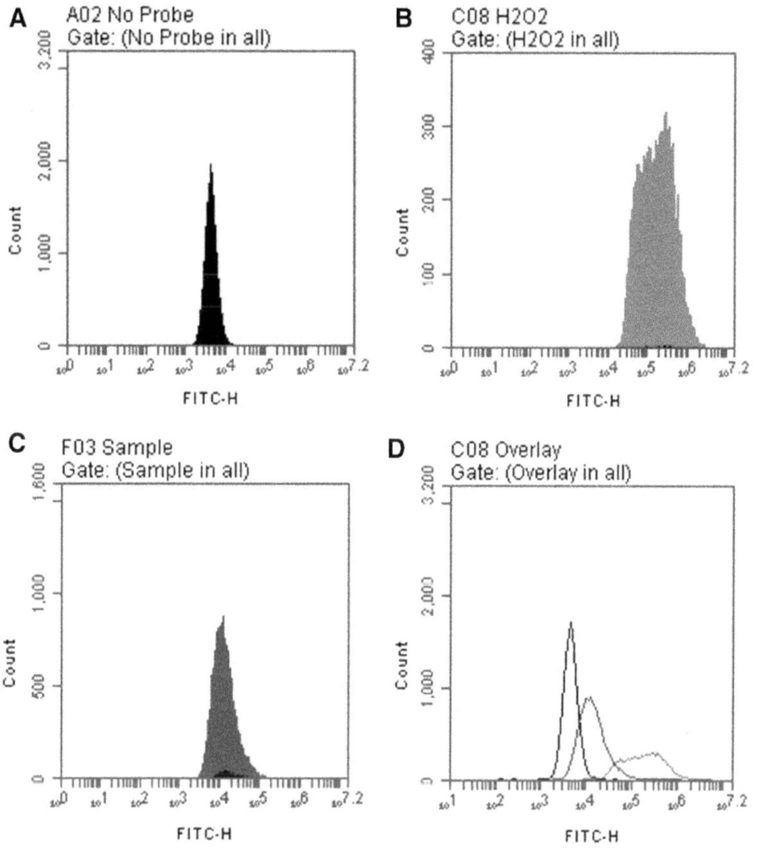

Fig. 3 Histograms of BD Accuri C6 following addition of CM-H2DCFDA. (**A**) No probe control, whereby differentiated myotubes' baseline autofluorescence was detected on the FL-1 channel (FITC). (**B**) Positive control whereby 50% v/v H_2O_2 was added to the cells in addition to CM-H2DCFDA for 5 min, which greatly increases fluorescent intensity. (**C**) Example of a standard C2C12 myotube sample. (**D**) Overlay channel demonstrating where a standard sample typically fits between the negative and positive controls

5. The amount of CM-H2DCFDA solution to resuspend cells in will depend on total cell number. Higher cell densities can cause instrumental blockage, so 500–1000 μL for 1–2 wells of a 6-well plate is an appropriate starting amount (this may need titrating depending on your experiment).
6. Cells should then be incubated in CM-H2DCFDA solution at 37°C for 15 min before subsequent analysis using flow cytometry.
7. A no-probe control should be used to determine baseline cellular autofluorescence, alongside a positive control whereby cells should be briefly suspended for 5 min in 50% v/v H_2O_2 in addition to CM-H2DCFDA to demonstrate efficacy of the probe (*see* **Note 8**).

8. Measurement of general ROS concentrations via CM-H2DCFDA can be determined using the FL-1 channel on the BD Accuri C6 (although the specific channel may differ depending on instrument used) (see Fig. 3).

3.2.2 Mitochondrial ROS Staining

1. Mitochondrial ROS can be quantified via flow cytometry using the fluorescent ROS probe MitoSOX Red. MitoSOX Red localizes in the mitochondria and can be oxidized by ROS, which increases its fluorescence intensity, allowing for quantification of mitochondrial-specific ROS. The excitation/emission maxima of the MitoSOX Red probe are 510/580 nm (*see* **Note 9**).

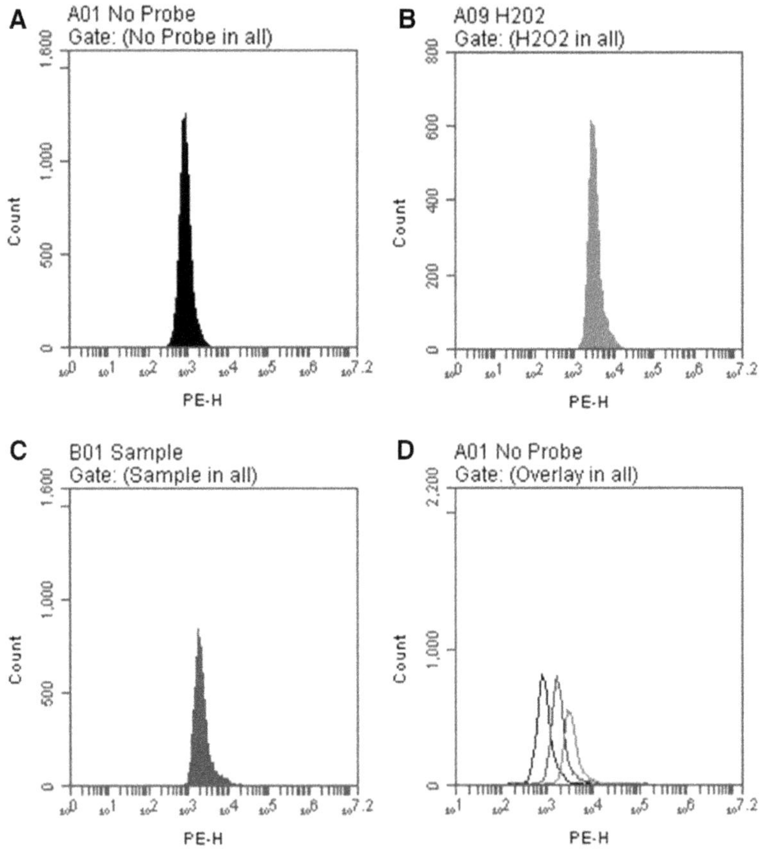

Fig. 4 Histograms of from a BD Accuri C6 following addition of MitoSOX Red. (**A**) No probe control displaying differentiated myotubes baseline autofluorescence as detected on the FL-2 channel (PE). (**B**) Positive control whereby 50% v/v H_2O_2 was added to the cells in addition to MitoSOX Red for 5 min which increases fluorescent intensity. (**C**) Example of a standard C2C12 myotube sample. (**D**) Overlay channel demonstrating where a standard sample typically fits between the negative and positive controls

2. Remove cell medium and wash twice in PBS before adding trypsin for 5 min.
3. Neutralize trypsin using DM and centrifuge cells at 300 rcf for 5 min.
4. Remove excess trypsin/media supernatant and resuspend the cell pellet in MitoSOX Red solution (5 μM MitoSOX Red in PBS) and incubate at 37°C for 15 min (*see* **Note** 7).
5. As with the CM-H2DCFDA, 500–1000 μL resuspension is appropriate for 1–2 wells of a 6-well plate to avoid instrument blockage; however, this may need adjusting depending on the instrument used.
6. Vortex the Eppendorf tubes briefly before analysis.
7. A negative control with control cells without MitoSOX Red administration alongside a positive control whereby 50% v/v H_2O_2 should be added alongside MitoSOX Red for 5 min (*see* **Note 8**).
8. Analysis was performed on the FL-2 channel of the BD Accuri C6, but this may differ depending on the instrument used (Fig. 4).

4 Notes

1. The FSC:SSC profile will depend on the rate of cellular differentiation. Given that 100% differentiation does not occur, be mindful that the scatterplot may differ between experiments. Gating should be tailored for each experiment to ensure cell populations are being appropriately characterized.
2. The FSC:SSC logarithmic plot displayed in Figs. 1 and 2 has been zoomed in for improved clarity. If you are unsure which population the cells in the logarithmic gate correspond to relative to the linear count gate, then you can back gate. Simply gate around cells within the logarithmic gate and make a new linear count panel. Collect the cells from those contained within the newly gated logarithmic panel, and this will correspond to whichever cell population is contained within the gate.
3. Wash steps should be utilized between experimental conditions, especially if you are also using dyes such as propidium iodide/Annexin V, which could stain subsequent cells. Washes can be performed briefly (10 s) with 5% bleach or ethanol.
4. The density of cells within the solution will influence the likelihood of cells blocking the instrument. While myotubes can be run through the flow cytometer, they are still large cells

relative to other cell types. As such, care should be taken to avoid instrument blockage. Diluting cells in larger amounts of PBS solution can help reduce the cell density and minimize the likelihood of blockage.

5. Cytometers will have options which allow for the tailoring of flow speed/core size. The higher the speed, the larger the core size. While fast speeds will increase the coefficient of variance slightly (due to more sample passing through the detectors at once), it will also increase the core size. Therefore, if repeated blockages are occurring, then running at a faster speed may be beneficial.
6. The CM-H2DCFDA probe is a chloromethyl derivative of the probe H2DCFDA, which provides better live cell retention when compared to H2DCFDA [15]. CM-H2DCFDA is a cell-permeable, non-fluorescent agent. Once CM-H2DCFDA diffuses across the cell membrane, the acetate groups of the compound are removed by cellular esterases to form the non-fluorescent reduced dye H2-DCF. The oxidation of H2-DCF results in the production of the fluorescent product DCF. Subsequently, increases in this ROS-dependent fluorescent product are, therefore, proportional to increases in cellular ROS concentrations.
7. CM-H2DCFDA and MitoSOX Red are both light-sensitive probes and care should be taken to minimize light exposure. Additionally, if obtained in large quantities, they should be aliquoted in foil to reduce freeze–thawing.
8. Positive controls do not have to be 50% v/v H_2O_2 and can be other compounds which will cause increases in cellular ROS concentrations, including antioxidant inhibitors (e.g., RSL3).
9. Hydroethidine (HE) is a two-electron reduced form of ethidium (Etd+) and can be readily taken up and internalized by live cells [16]. HE is the reduced form of Etd+ and can be oxidized by ROS, after which it intercalates into nucleic acids, increasing its fluorescent intensity [17]. MitoSOX Red is comprised of a HE covalently bonded to a positively charged hexyl triphenylphosphonium (TPP+) cation. TPP+ is often used for quantification of mitochondrial-specific products as the positive charge on the phosphonium group can facilitate transport across the outer and inner mitochondrial membranes, allowing for its accumulation in the negatively charged mitochondrial matrix [18]. However, MitoSOX Red is often marketed as a superoxide ($O^{2\bullet -}$) specific indicator, and while $O^{2\bullet -}$ is likely to be a major ROS oxidizing MitoSOX Red within the mitochondrial compartment, other ROS can also target MitoSOX Red. Therefore, quantification of mitochondrial superoxide requires further addition of superoxide-specific scavengers [19].

References

1. Pérez-Schindler J, Handschin C (2019) Chapter 11 - Physiological regulation of skeletal muscle mass: resistance exercise-mediated muscle hypertrophy. In: Walrand S (Ed.) Nutrition and skeletal muscle. Academic Press, London, United Kingdom: 125 London Wall, London, EC2Y 5AS, UK, 139–150
2. McCuller C, Jessu R, Callahan AL (2024) Physiology, skeletal muscle. StatPearls Publishing, Treasure Island, FL
3. Hargreaves M, Spriet LL (2020) Skeletal muscle energy metabolism during exercise. Nat Metab 2(9):817–828. https://doi.org/10.1038/s42255-020-0251-4
4. Vainshtein A, Sandri M (2020) Signaling pathways that control muscle mass. Int J Mol Sci 21(13):4759. https://doi.org/10.3390/ijms21134759
5. Yu BP (1994) Cellular defenses against damage from reactive oxygen species. Physiol Rev 74(1):139–162. https://doi.org/10.1152/physrev.1994.74.1.139
6. Murphy MP (2009) How mitochondria produce reactive oxygen species. Biochem J 417(1):1–13. https://doi.org/10.1042/bj20081386
7. Hempel N, Trebak M (2017) Crosstalk between calcium and reactive oxygen species signaling in cancer. Cell Calcium 63:70–96. https://doi.org/10.1016/j.ceca.2017.01.007
8. Brand MD (2010) The sites and topology of mitochondrial superoxide production. Exp Gerontol 45(7-8):466–472. https://doi.org/10.1016/j.exger.2010.01.003
9. Henríquez-Olguin C et al (2019) Cytosolic ROS production by NADPH oxidase 2 regulates muscle glucose uptake during exercise. Nat Commun 10(1):4623. https://doi.org/10.1038/s41467-019-12523-9
10. Okutsu M et al (2014) Extracellular superoxide dismutase ameliorates skeletal muscle abnormalities, cachexia, and exercise intolerance in mice with congestive heart failure. Circ Heart Fail 7(3):519–530. https://doi.org/10.1161/circheartfailure.113.000841
11. Bienert GP, Chaumont F (2014) Aquaporin-facilitated transmembrane diffusion of hydrogen peroxide. Biochim Biophys Acta 1840(5):1596–1604. https://doi.org/10.1016/j.bbagen.2013.09.017
12. Yin H, Xu L, Porter NA (2011) Free radical lipid peroxidation: mechanisms and analysis. Chem Rev 111(10):5944–5972. https://doi.org/10.1021/cr200084z
13. Pizzino G et al (2017) Oxidative stress: harms and benefits for human health. Oxid Med Cell Longev 2017:8416763. https://doi.org/10.1155/2017/8416763
14. Nolan A et al (2024) Fluorescent characterization of differentiated myotubes using flow cytometry. Cytometry A 105(5):332–344. https://doi.org/10.1002/cyto.a.24822
15. Ameziane-El-Hassani R, Dupuy C (2013) Detection of intracellular reactive oxygen species (CM-H2DCFDA). Bio-protocol 3(1):e313. https://doi.org/10.21769/BioProtoc.313
16. Gallop PM, Paz MA, Henson E, Latt SA (1984) Dynamic approaches to the delivery of reporter reagents into living cells. Biotechniques 2(1):32–36.
17. Bucana C, Saiki I, Nayar R (1986) Uptake and accumulation of the vital dye hydroethidine in neoplastic cells. J Histochem Cytochem 34(9):1109–1115. https://doi.org/10.1177/34.9.2426339
18. Ross MF et al (2005) Lipophilic triphenylphosphonium cations as tools in mitochondrial bioenergetics and free radical biology. Biochemistry (Mosc) 70(2):222–230. https://doi.org/10.1007/s10541-005-0104-5
19. Kauffman ME et al (2016) MitoSOX-based flow cytometry for detecting mitochondrial ROS. React Oxyg Species (Apex) 2(5):361–370. https://doi.org/10.20455/ros.2016.865

Chapter 14

Measuring Skeletal Muscle Mitochondrial Architecture in 3D Using Focused Ion Beam Scanning Electron Microscopy (FIB-SEM)

Hailey A. Parry and Brian Glancy

Abstract

Recent advances in volume electron microscopy have permitted unprecedented visualizations and in-depth quantitative analyses of mitochondrial network structure in skeletal muscle. Nanoscale, 3D analyses of skeletal muscle mitochondria have provided novel insights into how mitochondria interact among themselves and with other skeletal muscle cell organelles, such as the sarcoplasmic reticulum and the sarcomere. The identification of mitochondria within volume electron microscopy datasets and the downstream analyses of segmented structures remain two major bottlenecks in the analysis of 3D mitochondrial network structure at the nanoscale. Here, we provide a step-by-step guide on how to use machine learning programs to segment skeletal muscle mitochondria from volume electron microscopy datasets, followed by instructions on several types of 3D analyses of individual mitochondria and mitochondrial network structure that can be performed using ImageJ. We highlight key quality-control steps to ensure precise and accurate segmentation and analyses and to enhance reproducibility across groups.

Keywords Mitochondria, Skeletal muscle, Volume electron microscopy, Focus Ion Beam Scanning Electron Microscopy

1 Introduction

Technological advances in volume electron microscopy have paved the way for our evolving understanding of mitochondrial networks. Skeletal muscle mitochondria were first described as a reticulum in 1966 using 2D transmission electron microscopy (TEM) [1]. The mitochondrial network was confirmed when recreating a 3D structure of skeletal muscle mitochondria using serial section TEM [2]. It was not until 1985 that scanning-electron microscopy (SEM) was performed on rat hindlimb skeletal muscle, demonstrating fiber-type-specific dependency of mitochondrial network connectivity [3]. Although highly informative, the pseudo-3D nature of the

Joe Quadrilatero (ed.), *Myogenesis and Skeletal Muscle Biology*, Methods in Molecular Biology, vol. 3057, https://doi.org/10.1007/978-1-0716-5396-8_14,

SEM images did not allow for quantification of the mitochondrial network. The Brooks laboratory used projection images to create a quantitative image to calculate the surface area and volume of mitochondria [4]. Despite these technological advances to create 3D images, the bottleneck for volume electron microscopy remained limited to technology, leading to minimal quantitative measurements on mitochondrial networks.

Advances over the last two decades in large-scale, 3D electron microscopy techniques, termed volume electron microscopy, now allow for capturing 3D images of subcellular structures in cells and tissues [5, 6]. One of several volume electron microscopy techniques, Focused Ion Beam-Scanning Electron Microscopy (FIB-SEM), uses SEM to provide a 2D image of the sample surface, followed by the use of an ion beam to cut a nanometer-scale layer (as small as 3 nm) off the top of the sample. These imaging and ion beam ablation steps are then repeated many times to create a stack of up to thousands of 2D images, which can then be combined into a 3D volume [6]. The use of FIB-SEM confirmed that human skeletal muscle mitochondria have a similar morphology and connectivity to rodent skeletal muscle mitochondria [7]. Additionally, FIB-SEM confirmed the observation of mitochondrial pooling around the cell's periphery to be associated with an embedded capillary [8]. Notably, the different pools of mitochondria were recently quantified using FIB-SEM, confirming that mitochondria next to a capillary, termed paravascular mitochondria, are in fact larger, more spherical, and in higher contact with neighboring mitochondria than interfibrillar mitochondria [9].

The development of FIB-SEM has opened the door to subcellular imaging for a deeper understanding of cellular organization. With the increase in imaging speed, the bottleneck for volume electron microscopy is now organelle segmentation. While the ability to visualize and quantify the sizes, shapes, and interactions among many different subcellular structures simultaneously is a major strength of electron microscopy, it also provides a significant challenge. The greyscale nature of the images means that each organelle has to be identified by its contrast, shape, and location rather than the organelle-specific labels often used in fluorescent light microscopy techniques. A trained image analyst can quickly identify and manually trace a mitochondrion within a single electron microscopy image for morphological analyses. However, this manual segmentation approach is not feasible for entire volume electron microscopy datasets with thousands of images and hundreds to thousands of mitochondria. Fortunately, over the last decade, many semi-automated segmentation programs using machine and deep-learning approaches have helped overcome the bottleneck of segmenting cellular structures [10–13]. Each segmentation method has tradeoffs; deep learning using a neural network is generally more powerful and better able to identify structures of interest within lower contrast images, but the requirements for training and running the models can be very time consuming. Conversely, simpler machine learning programs are less powerful but require lower training and run times, making these programs useful for datasets with clear contrast and easily identifiable structures.

This chapter will mainly focus on machine learning segmentation, quality control, and image analysis, which occur after FIB-SEM collection. The preparation steps provided have been optimized for a single mouse soleus skeletal muscle (*see* **Note 1**). Furthermore, the staining steps included herein provide optimal contrast for membranes. Readers are directed elsewhere for staining methods to highlight other cellular structures [14–16]. We highlight numerous quality control steps necessary to determine accurate data, most of which have yet to be published. We describe mitochondrial segmentation, but this protocol can be applied to any independent or continuous organelle. This step-by-step protocol will provide the expertise and insight to make it more widely available for scientists.

2 Materials

Use 7 mL glass vials when possible. Keep liquid waste separate to avoid chemical reactions. When in doubt, throw the liquid waste away with chemicals from the previous step. All materials in contact with osmium tetroxide should be washed three times with distilled water before being thrown away. Additionally, be cautious when handling radioactive materials and be sure the laboratory follows all regulations for storing and handling radioactive materials.

2.1 Sample Preparation—Day 1

1. Isoflurane.
2. In vivo fixative solution: 2% glutaraldehyde in 0.1 M phosphate buffer, pH 7.2.
3. Alto Cornol Brain Matrix slicing block (Roboz Surgical Store, MD, USA).
4. Slice-edged blade (Uline, WI, USA).
5. Heavy fixative solution: 2.5% glutaraldehyde, 1% formaldehyde, and 0.1 M sodium cacodylate, pH 7.2.
6. Sodium cacodylate wash solution: 0.1 M sodium cacodylate buffer, pH 7.2.
7. Ferricyanide and osmium staining solution: 1.5% ferricyanide, 2% osmium tetroxide, and 0.2 M sodium cacodylate, pH 7.2. For a final volume of 10 mL, dissolve 150 mg Ferricyanide in 5 mL 0.2 M sodium cacodylate buffer. Sonication can be used to speed up this process. Lastly, add 5 mL of 4% Osmium tetroxide to make a final concentration of 1.5% Ferricyanide and 2% Osmium tetroxide.

8. TCH solution: 1% TCH (thiocarbohydrazide) solution. Add 10 mL of distilled water with 100 mg of TCH to a glass vial. Place the glass vial in a 60°C oven. Swirl occasionally until dissolved. Once the solution is opaque pink, filter the solution through a 22-micrometer syringe filter and cool to room temperature (*see* **Note 2**).
9. Osmium tetroxide solution: 2% osmium tetroxide solution. For 10 mL, add 5 mL of distilled water and 5 mL of 4% osmium tetroxide to a 20 mL vial. Chill on ice before use.
10. Uranyl acetate solution: 1%–2% aqueous uranyl acetate in distilled water.

2.2 Sample Preparation—Day 2

1. Lead aspartate solution. Dissolve 39.9 mg of L-aspartic acid in 10 mL of distilled water. Place in a water bath and swirl occasionally to dissolve. Once dissolved, add 66 mg of lead nitrate to the solution and heat under stirring to 60°C. Adjust pH to 5.5 with 1 M NaOH. This solution is prepared first on day 2 of sample preparation (*see* **Note 3**).
2. 30% ethanol solution. Dilute ethanol to 30% concentration with distilled water.
3. 50% ethanol solution. Dilute ethanol to 50% concentration with distilled water.
4. 70% ethanol solution. Dilute ethanol to 70% concentration with distilled water.
5. 80% ethanol solution. Dilute ethanol to 80% concentration with distilled water.
6. 95% ethanol solution. Use a previously opened bottle of ethanol.
7. 100% ethanol solution. Use an unopened bottle of ethanol.
8. The Epon mixture (Embed 812 Embedding Kit, Electron Microscopy Sciences): 55.5% Embed-812 (Epon-812 substitute), 30.95% DDSA (dodecenylsuccinic anhydride), 31.25% NMA (nadic methyl anhydride), and 1.625% B-DMA (benzyl dimethyl amine) or DMP-30 (2,4,6-tri-(dimethylaminomethyl)phenol) weight/volume. It should be mixed with a stir bar in a vacuum-sealed container for a minimum of 1 h before use.
9. 2:1 ethanol: Epon mixture solution. For 15 mL, combine 10 mL of 100% ethanol with 5 mL of Epon mixture.
10. 1:1 ethanol: Epon mixture solution. For 10 mL, combine 5 mL of 100% ethanol with 5 mL of Epon mixture.

2.3 Sample Preparation—Day 3

1. The Epon mixture (Embed 812 Embedding Kit, Electron Microscopy Sciences): 55.5% Embed-812 (Epon-812 substitute), 30.95% DDSA (dodecenylsuccinic anhydride), 31.25% NMA (nadic methyl anhydride), and 1.625% B-DMA (benzyl

dimethyl amine) or DMP-30 (2,4,6-tri-(dimethylaminomethyl)phenol) weight/volume. It should be mixed with a stir bar in a vacuum-sealed container for a minimum of 1 h before use.

2. Aluminum SEM pin stubs (Electron Microscopy Sciences, Hatfield, PA). These pin stubs have a slotted head ½″ diameter, pin ⅛″ diameter, 6 mm height.
3. Emery paper. (Electron Microscopy Sciences, Hatfield, PA).

2.4 Sample Overview Images

1. Zeiss Crossbeam 540.
2. Zeiss Atlas 5 software (Carl Zeiss Microscopy, GmbH, Jena, Germany).

2.5 Segmentation of Cellular Structures

1. Image J (National Institutes of Health, Bethesda, MD, USA).
2. Ilastik (www.ilastik.org).

3 Methods

3.1 Sample Preparations—Day 1

1. Anesthetize the mouse with 2%–5% isoflurane via a nose cone. Peel the skin of the hindlimb back, and place the hindlimb into the in vivo fixative solution for 30 min. Remove the entire soleus and euthanize the animal via exsanguination.
2. Place the soleus into an Alto Cornol Brain Matrix slicing block and slice into 1 mm thick sections using a slice-edged blade. Place sections into the heavy fixative solution and incubate at room temperature for 60–90 min. If needed, leave samples overnight.
3. Wash samples five times for 3 min each in sodium cacodylate wash solution.
4. Stain samples with ferricyanide and osmium staining solution on ice in the dark for 1 h.
5. Confirm that samples appear black. Wash samples five times for 3 min each in distilled water.
6. Stain samples with TCH solution for 20 min at room temperature on a rocker.
7. Wash samples five times for 3 min each in distilled water.
8. Stain samples with osmium tetroxide solution for 30 min on ice in the dark.
9. Wash samples five times for 3 min each in distilled water.
10. Leave samples in 1 mL uranyl acetate solution at 4°C overnight. Set aside 100 mL of distilled water at 4°C overnight and another 100 mL of distilled water at 60°C overnight.

3.2 Sample Preparation—Day 2

1. Wash samples five times for 3 min each in chilled distilled water. After the final wash, incubate samples at 60°C for 10 min.
2. Stain samples with lead aspartate solution for 30 min at 60°C.
3. Wash samples five times for 3 min each in warmed distilled water.
4. Dehydrate samples in increasing concentrations of ethanol: 30%, 50%, 70%, 80%, 90%, 95%, and 100%. Perform each concentration step twice in succession (e.g., 30% then 30%), incubating for 5 min each time.
5. Incubate samples at room temperature for 1 h in a 2:1 ethanol: Epon mixture.
6. Leave samples in a sealed vacuum hood overnight in a 1:1 ethanol: Epon mixture. Loosen the vial caps to allow ethanol to evaporate.

3.3 Sample Preparation—Day 3

1. Incubate the sample for 3 h in the Epon mixture. For the final wash, place the samples in a vacuum-sealed container with the caps off.
2. Prepare stubs for sample mounting by rubbing the top of each aluminum stub with Emery paper to create a rough surface. Wash stubs with acetone, then with water. Place cleaned stubs on a Kimwipe to dry.
3. Mount samples on stubs by adding a small drop of Epon mixture to the center of the aluminum stub. Place the sample in the Epon mixture, remove excess Epon mixture with filter paper, and position the sample in the center of the stub, avoiding the edges.
4. Place mounted samples in a 60°C oven under vacuum for 2 days.

3.4 Sample Overview Images

1. Image mounted samples with a Zeiss Crossbeam 540 at low resolution to obtain overview images of prepared samples. Export images from the Zeiss Atlas 5 software as TIFF files.
2. Select a region of interest in one of the overview images to perform high-resolution 3D imaging (*see* **Note 4**).

3.5 FIB-SEM Imaging

1. Acquire images of the selected cell with a Zeiss Crossbeam 540 using the Zeiss Atlas 5 software. Use an in-column energy-selective backscatter with a filtering grid during image collection to reject unwanted secondary electrons and backscatter electrons, up to a voltage of 1.5 kV at a working distance of 5.01 nm.
2. Mill the sample with a focused ion beam (FIB) at 30 kV and 2–2.5 nA beam current. Set slice thickness to 5–10 nm to achieve a total depth of 10–20 μm.
3. Align images using the proprietary algorithm in the Atlas 5 software. Export image stacks in TIFF format for future analysis in ImageJ.

3.6 Segmentation of Cellular Structures (see Note 5)

1. Load raw FIB-SIM images into ImageJ (National Institutes of Health, Bethesda, MD, USA). Apply the CLAHE plug-in to enhance image contrast. Export images as a single HDF5 file using the Ilastik plugin for ImageJ. Save raw images as TIFF files for future access (*see* **Note 6**).
2. Open Ilastik and start a new "Pixel Classification" project. Load the saved HDF5 file into the "Raw Data" section. Move to "2. Feature Selection" and select all features for color/intensity, edge, and texture. Move to "3. Training" and draw labels for cellular structures (Fig. 1A). Draw an outer membrane label (yellow) separately from the inside (blue) for individual structures, such as mitochondria or lipid droplets. Include an "other" label (red) for background or unwanted regions (Fig. 1B). Select "Live Update" to train the model. After the update completes, turn off "Live Update" and correct labels as needed. Train images in the beginning, middle, and end planes

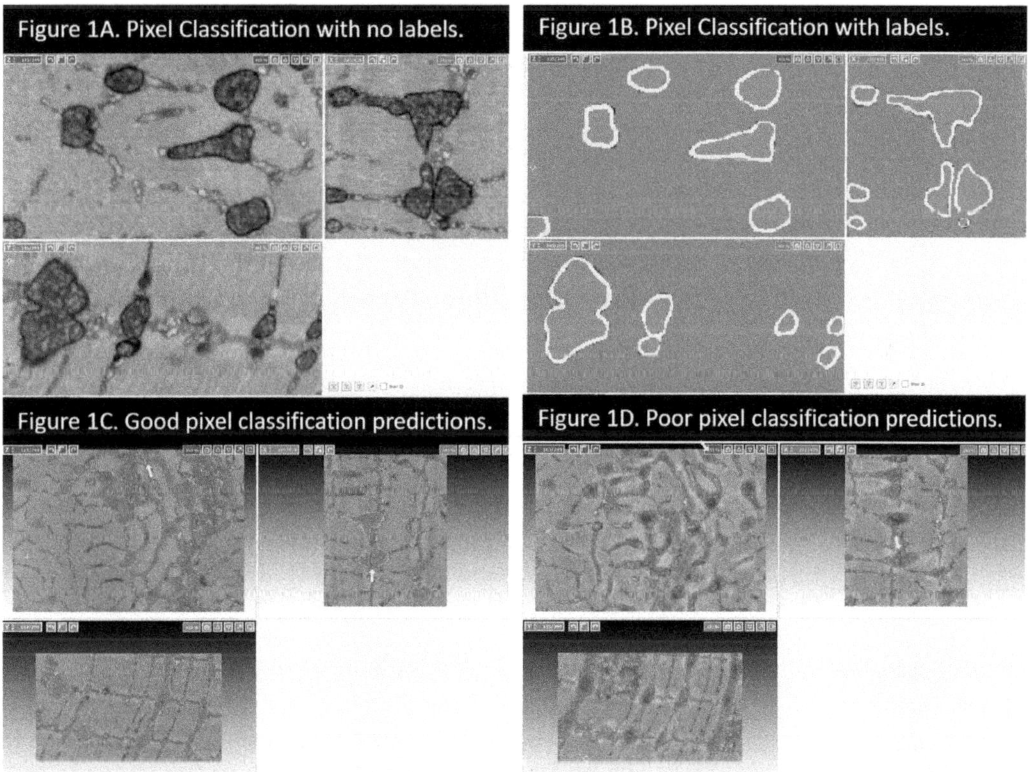

Fig. 1 Pixel classification of mitochondria. To classify each pixel of the raw gray image set, load the raw data as an HDF5 file (**A**). Create a label for "outer mitochondria" (yellow), "inner mitochondria" (blue), and "other" (red). Select an appropriate amount of pixels to train the model on the correct boundaries of the mitochondria (**B**). This should be done at each data set's beginning, middle, and end in each plane. After selecting "Live Update," the model will learn how to classify each pixel. Good pixel classification has predictions with crisp boundaries (white arrow) that can separate close but distinct mitochondria (yellow arrow; **C**). Poor pixel classification results in hazy outlines, indicative of model uncertainty (**D**)

of the stack. Test the model on untrained slices and increase the "Prediction" labels to assess accuracy (Fig. 1C). When mitochondria are defined from predictions in all three planes, then the model is trained well enough (*see* **Note 7**).

3. Move to "4. Prediction Export." In "Export Settings," select "Choose Export Image Settings." Export predictions for the outer membrane of the organelle of interest. Ensure all z, y, and x ranges are selected. Convert the export to floating 32-bit format and renormalize the data range to 0–255.
4. Create a new "Boundary-based Segmentation with Multicut" project in Ilastik. Load the HDF5 raw images from the Pixel Classification project. In the "Probabilities" tab, load the exported HDF5 file. Move to "2. Watershed" and adjust the threshold to highlight all outer mitochondria with minimal noise. Select "Update Watershed" to create superpixels (Fig. 2). If organelle boundaries are unclear, return to "2. Watershed" or retrain Pixel Classification to improve boundary predictions (*see* **Note 8**).
5. Train organelle segmentation by turning raw data superpixels (yellow lines) into labels. Right-click organelle outlines to highlight in red, and left-click to turn remaining yellow lines green (Fig. 3B). Repeat this process on multiple sections until sufficient "mito" and "other" labels are applied (red and green arrows in Fig. 3B). Select "Update Now" and verify predictions: mitochondria boundaries should appear in blue, with mostly yellow lines outside and inside the mitochondria (Fig. 3C). Correct splitting errors, if necessary. Focus training on intermitochondrial junctions and perform training in beginning, middle, and end planes of the image stack. Select "Live Multicut" with a beta value of 0.3–0.5 to generate segmentation. Export files after achieving approximately 95% accuracy (*see* **Note 9**).
6. Export data in the "4. Data Export" tool. Select "Choose Export Image Settings…," name the file, and select the location. Click "OK" and select "Export" under the "Action" label to save the file.

3.7 Cleaning and Quality Control of Segmentation

1. Import the exported Multicut HDF5 file into ImageJ using the Ilastik plug-in. Save the file as a TIFF. Load the FIB-SIM raw data into ImageJ. Merge the segmented TIFF file with the raw data under "Image" and "Color."
2. Review the entire overlaid image stack for accurate and precise predictions of mitochondria and organelles. Correct general size errors using the "Morphological Filters (3D)" tool in the MorphoLibJ plug-in if the entire organelle is consistently 2 pixels too large or too small at every angle (Fig. 4A and B;

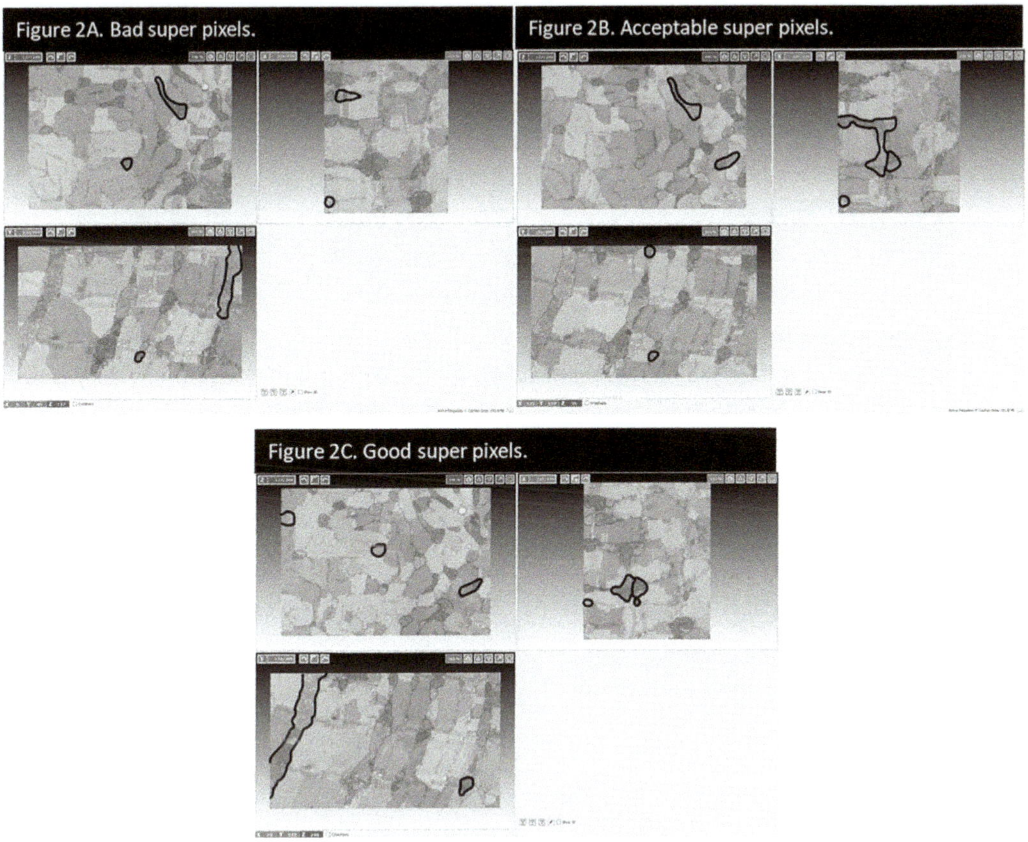

Fig. 2 Creating super pixels. Super pixels should define an individual mitochondrion with separate colors from the environment. **(A)** is an example of bad super pixel predictions. The mitochondria are outlined in black. Notice how the colors inside the outlined mitochondria extend to the surrounding areas. This will result in poor boundary predictions in the next step. Acceptable super pixel generation (**B**) is when mitochondria are separated from the surrounding environment, but many colors exist within one mitochondrion's boundary. This will lead to more work to define proper boundaries. Good super pixel generation (**C**) is when mitochondria are identified from the rest of the image with minimal breaks (aka colors) within mitochondria

see **Note 10**). Select "dilate" to enlarge labels and "erode" to shrink labels.

3. Check for split errors (Fig. 4A). Merge mitochondria that are incorrectly split by selecting the exported mitochondrial segmentation TIFF file, opening "Label Images" → "Label Edition" in MorphoLibJ, selecting labels to merge, and clicking "Merge." Delete labels for one or two minor merging errors. If merging errors are frequent, adjust the watershed in the Multicut project in Ilastik (Sect. 3.7). Verify segmentation: each mitochondrion should have its own label, perfectly overlaying the raw data (Fig. 4C). Refer to the final mitochondrial segmentation example (Fig. 5).

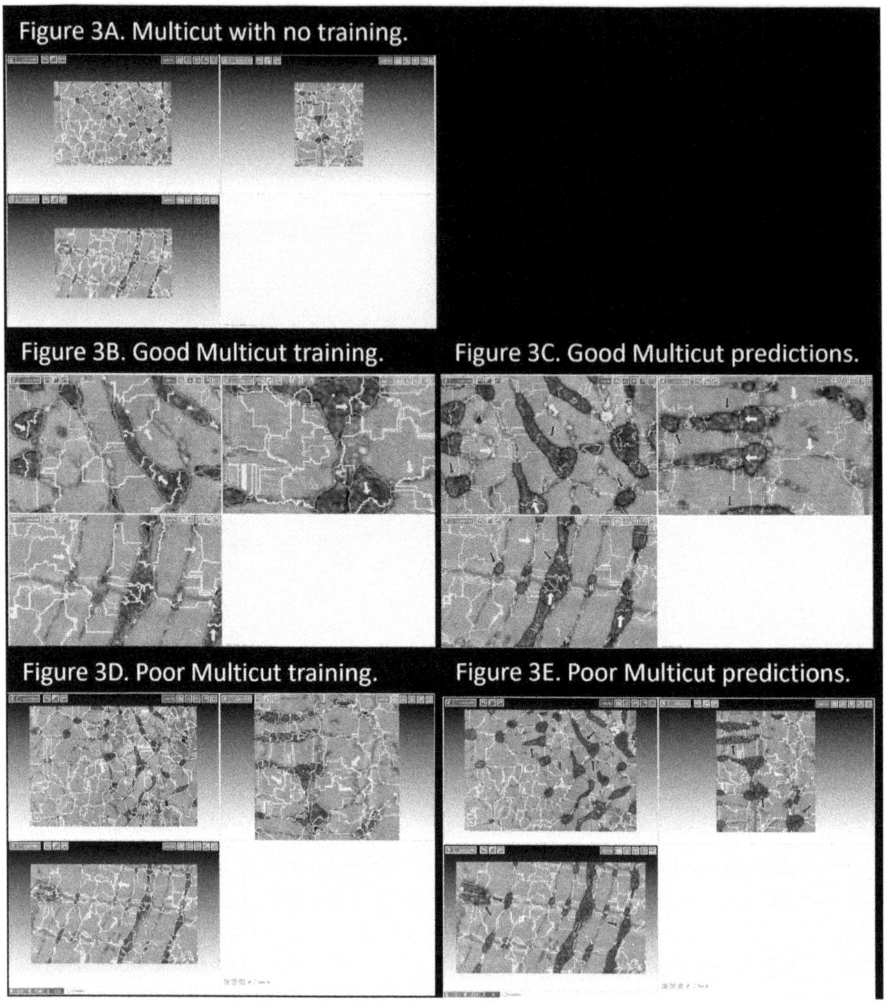

Fig. 3 Multicut training. The goal of multicut training is to define the boundaries of the mitochondria. After loading the raw data with the outer mitochondrial membrane probabilities, the dataset will have yellow lines that outline the superpixels generated in the previous step (**A**). Good multicut training will properly identify the boundaries of the mitochondria in red while highlighting the rest of the yellow lines as green (**B**). This should result in good boundary predictions, where the boundaries of the mitochondria are selected in blue (**C**). Poor multicut training happens when little guidance is provided regarding what is or is not a mitochondrial boundary (**D**). When too little input is provided, a poor multicut prediction will occur (**E**). This is defined as one mitochondrion being defined as multiple mitochondria or partial identification of a mitochondrial boundary

3.8 Individual Mitochondria Analysis

1. Load the 32-bit mitochondria labels into ImageJ. Duplicate the mitochondrial label image. Create a binary image by selecting all mitochondria with the "Manual Threshold" tool. Apply the "Make Binary" function under Process → Filters. Perform the "Histogram" function on the binary image. Select the "List" option in the histogram window. Record the number of counts at 255 as the total mitochondrial pixels ("N").

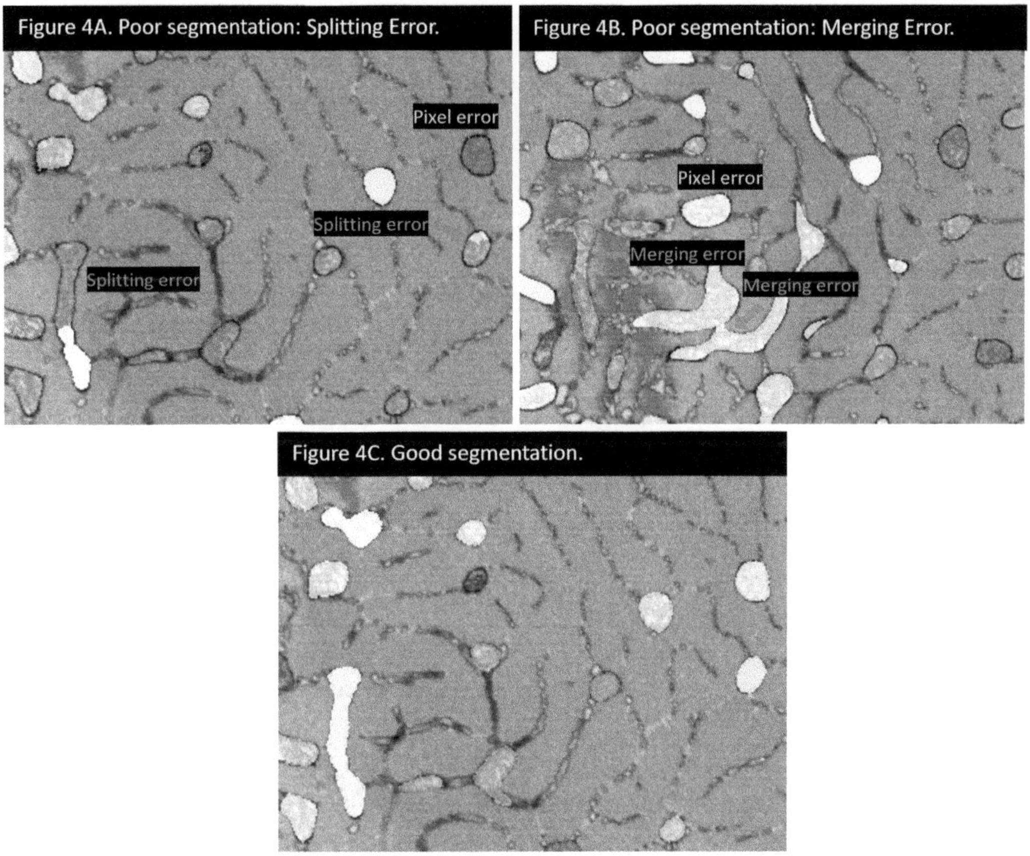

Fig. 4 Quality control. To determine if the multicut training produced appropriate mitochondrial segmentation, merge the raw data files and the segmented mitochondrial labels in ImageJ. Poor segmentation can be seen with a high number of splitting errors (**A**), merging errors (**B**), or pixel errors (**A** and **B**). Splitting errors are when one mitochondrion is identified as two or more. These can be fixed within ImageJ. Merging errors are when multiple mitochondria have been defined as one mitochondrion. This is much harder to fix, and leaning on the slide of splitting errors is suggested. Pixel error may occur if the segmentation is smaller than the actual mitochondria. Filters can correct this as long as all mitochondria are smaller or larger by the same pixel value. An example of good segmentation with no splitting, merging, or pixel errors is shown in panel **C**

2. Measure mitochondrial properties using MorphoLibJ: select "Analyze Regions 3D" to obtain Voxel Count, Volume, Surface Area, and Sphericity. Use "Geodesic Diameter 3D" in the Analyzed section to measure mitochondrial length [17]. Save all data as a .csv file.
3. Verify individual mitochondrial measurements: randomly select a mitochondrion from the "Analyze Regions 3D" output. Use the "Select Labels" tool in MorphoLibJ to isolate that mitochondrion in a new image stack. Measure the length, width, and number of slices using the line tool to calculate approximate volume. Compare this manual calculation with

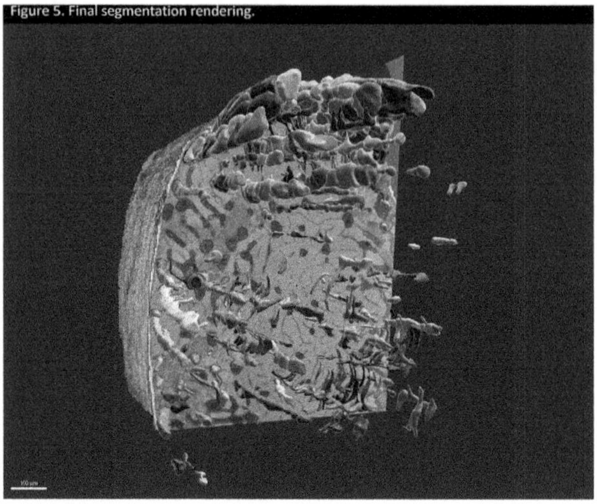

Fig. 5 Final segmentation. This is an example of the mitochondria segmentation overlaid with the raw data in 3D. A different color identifies each mitochondrion

the pixel-based volume from MorphoLibJ. Apply the same approach to validate surface area and sphericity.

4. Open the .csv files in Microsoft Excel. Convert voxel measurements to appropriate units based on image resolution. For 10 nm × 10 nm × 10 nm FIB-SIM images, multiply voxel counts by 0.001 for volume, 0.01 for surface area, and 0.1 for geodesic diameter (length). Leave sphericity values unchanged, as they are unitless.

3.9 Mitochondria to Organelle Interactions

1. Load the 32-bit mitochondria labels into ImageJ. Create outer boundaries using "Label Boundaries" under Label Images in MorphoLibJ. Divide the boundary image by itself using the Image Calculator to generate an image with 1.0 values for outer mitochondrial membrane pixels and NA for all others.
2. Load the segmented organelle of interest. Convert it into an 8-bit binary image. Duplicate the image stack. Apply the "Distance Transform 3D" tool under "Process" in ImageJ to generate a distance map, where the organelle center is 0 pixels. Use "Intensity Measurements 2D/3D" in MorphoLibJ to measure the minimum distance of mitochondria from the organelle. Use the distance map as input and the original mitochondrial labels as reference.
3. Select the mitochondrial label stack. Use "Assign Labels" in Label Images (MorphoLibJ) to assign minimum distance labels. Apply a 40 nm threshold using "Manual Threshold" to select membranes that form contact sites. Convert selected labels to binary using "Make Binary." Divide the contact site

image stack by itself and multiply by the original mitochondrial labels to isolate contact site membranes with correct mitochondrial labels.

4. Subtract the contact site membrane labels from the original mitochondrial labels using ImageJ's Image Calculator. Measure total voxels in each label with "Analyze Regions 3D" in MorphoLibJ. Save the results as a .csv file.
5. Calculate the percentage of the outer mitochondrial membrane in contact with the organelle: subtract voxel values from Step 3.9 (part 4) from Step 3.8 (part 1), divide by the value from Step 3.8 (part 1), and multiply by 100.
6. Verify organelle connectomics calculations. Identify errors by checking for negative percentage values. Select individual mitochondria using "Select Labels" in MorphoLibJ and include the organelle of interest. Repeat the workflow above to determine which step causes the error.

3.10 Mitochondria-to-Mitochondria Interaction

1. Load the mitochondrial label image stack into ImageJ. Duplicate the stack and create a binary image. Divide the binary mitochondrial label stack by itself to produce a 32-bit (float) image with 1.0 values for mitochondrial pixels and NA for all other pixels. Multiply this binary NA image by the original mitochondrial label stack using the Image Calculator to generate an image of mitochondrial labels with NA background. Apply a $1 \times 1 \times 1$ pixel Mean 3D filter (Process → Filters → Mean 3D...) to adjust pixel values where adjacent pixels differ. Use the Image Calculator to perform a "difference" operation between the mean-filtered image and the original mitochondrial labels. Apply the "Absolute" function to convert any negative values to positive.
2. Create a binary image (0 and 255 pixel values) from the difference image. Subtract this binary difference image from the binary mitochondrial labels. Divide the result by itself and multiply by the original mitochondrial labels. Measure the number of voxels in the resulting mitochondrial labels using "Analyze Regions 3D" in MorphoLibJ. Save the output as a .csv file.
3. Calculate the percentage of the outer mitochondrial membrane in contact with neighboring mitochondria: subtract voxel values from Step 3.10 (part 2) from the values in Step 3.8 (part 1), divide by the value from Step 3.8 (part 1), and multiply by 100.
4. For quality control, perform the same steps described in Step 3.9 (part 6).

3.11 Mitochondrial Network Analysis

1. Rotate the binary mitochondrial labels until the skeletal muscle contraction axis is perfectly horizontal. Record the rotation

applied to the raw images so that z-disks are vertically aligned in both XY and XZ planes. Use the "Rotate" and "Reslice" tools in ImageJ. Apply the same rotations to the binary mitochondrial labels.

2. Use "Z Project" in Stacks to create an average projection image of the XY and XZ planes. Duplicate this image and convert it to a binary image with 0 and 255 values.
3. Apply the "OrientationJ Distribution" plugin to determine mitochondrial network angles [18]. Save the output as a .csv file. Define mitochondria as parallel if angles are between ± 0–10° and perpendicular if angles are between ±80° and 90°.
4. Calculate the average values in the "distribution of orientation" column for parallel and perpendicular mitochondria. Compute the ratio of parallel to perpendicular mitochondria to quantify network orientation.
5. Determine mitochondrial subnetworks using the "Particle Analyser" in the BoneJ plugin on the binarized mitochondrial dataset [19]. Uncheck all boxes except "Show particle stack." Record the volume of each subnetwork from the Results table.
6. Determine the number of mitochondria per subnetwork: load the 32-bit individual mitochondrial labels file. Divide the labels by the total number of mitochondria using Process → Math → Divide. Add the resultant image to the Subnetwork particle stack image using the Image Calculator, selecting 32-bit (float) as the output. Create a histogram (Analyze) of the resulting image, setting the number of bins equal to the number of mitochondria multiplied by the number of subnetworks. Copy histogram values into Microsoft Excel. Count non-zero values in the second column corresponding to first-column values between 1.0 and less than 2.0 to determine mitochondria in the first subnetwork. Repeat for each subnetwork.

4 Notes

1. This protocol is optimized for a single mouse soleus. The volume of solutions may have to be adjusted depending on tissue size.
2. Make this at the start of Step 3.1.4 for the best contrast in the images.
3. If precipitation occurs after pH adjustment, the lead is effectively removed from the solution. The solution must be remade if this happens. Precipitation occurs because the pH is above

5.5. One tip to avoid this is to count how many drops of 1 M NaOH are needed before precipitation and use this as a standard dropwise amount.

4. It is recommended to choose areas with undamaged cells at physiological sarcomere lengths (>1.6 μm) with high contrast. Cellular mitochondrial content within a cell can provide an index of the metabolic fiber type, and cell size may provide a marker of contractile fiber type when choosing skeletal muscle cell types.
5. Organelle segmentation is the bottleneck of this protocol. Sample preparation and image capture are expected to take 2 weeks, at minimum. The following segmentation and analysis can take anywhere from a week to a month, depending on the segmentation software. Post-image segmentation and analysis can be hastened if the original images have a sharp and clear resolution. If the image resolution is unclear, segmentation will be difficult regardless of the software package used.
   a. Computers with high RAM are needed for post-image segmentation and analysis using the Ilastik program as described below. We use a 2 TB RAM computer for datasets of ~4000 × 3000 × 2000 pixels. Smaller image stacks (i.e., 500 × 500 × 500 pixels) can be performed on a 64 GB RAM computer.
6. The segmentation steps described below use the software Ilastik [20]. Ilastik uses machine learning to identify objects and segment them quickly. This is best used when images have high contrast resolution. Other software packages using deep learning models can help identify objects when there is less contrast [11, 21]. However, these tools typically take longer to train. Importantly, in Ilastik, independent structures, such as mitochondria and lipid droplets, require pixel classification and a watershed step. Continuous structures, like the sarcotubular network (SR/T), only require pixel classification. Many tools are available to perform these steps.
7. An example of a not well-trained model is shown in Fig. 1D. The white arrow in Fig. 1C shows the clean probability prediction of the outer mitochondrial membrane. Notice the white arrow in Fig. 1D showing a blurred yellow line for the probability prediction of the outer mitochondrial membrane. This suggests the model is not trained enough to define the boundary. The yellow arrow in Fig. 1C shows a well-trained model that can predict two different mitochondria next to one another. The yellow arrow in Fig. 1D shows two mitochondria that are not predicted to be separated.

8. Good predictions should have lots of blue outlines on the boundary of the mitochondria (see blue arrows) with mostly yellow lines outside and inside the mitochondria (yellow arrows). There will be a few blue lines in the middle of a mitochondrion. This will result in a splitting error when one mitochondrion gets marked as multiple mitochondria. This can be corrected in future steps, but too many corrections can lead to more time spent in the quality control step. Often, only a few selections are needed to get the training started. It is easier to start with a few selections, correct the models' errors, and repeat the process. Figure 3D shows poor, minimal training as indicated by few red and green selections. This resulted in Fig. 3E, which had decent predictions of the outer mitochondria but high splitting predictions (blue lines in the middle of mitochondria). Additional training would focus on the lines located inside one mitochondrion to avoid future splitting errors. The most critical regions to have trained are the intermitochondrial junctions, as this is a key measurement needed for mitochondrial network analysis. Do this in each plane at the image stack's beginning, middle, and end. Next, select "Live Multicult" using a beta value between 0.3 and 0.5 to generate segmentation (Fig. 3C).
9. In Fig. 2A, an example of bad superpixels is shown. The outline of the mitochondria is provided in black. Notice that the colors inside the mitochondria do not create the mitochondrial shape. In Fig. 2B and C, the mitochondria have their own colors to create a distinctive shape. Figure 2B is labeled as "acceptable super pixels" because the mitochondria are properly separated from the rest of the pixels, but there are many super pixels (aka colors) within a mitochondrion. Figure 2C are "good super pixels" because the mitochondria are distinct and there are fewer super pixels within a mitochondria.
10. The instructions advise dilating/eroding if the mitochondria are off by 2 pixels. However, 2 pixels is an example. Mitochondria can be off by any number of pixels. The dilate/erode function will only work in this step if the mitochondria are consistently too large or too small in all directions. If the mitochondria are too long in the Y plane and too short in the X plane, then new segmentation must be completed.

References

1. Bubenzer H-J (1966) Die dünnen und die dicken Muskelfasern des Zwerchfells der Ratte. Z Zellforsch Mikrosk Anat 69:520–550
2. Bakeeva L, Chentsov YS, Skulachev V (1978) Mitochondrial framework (reticulum mitochondriale) in rat diaphragm muscle. Biochim Biophys Acta 501(3):349–369
3. Ogata T, Yamasaki Y (1985) Scanning electron-microscopic studies on the three-dimensional structure of mitochondria in the

mammalian red, white and intermediate muscle fibers. Cell Tissue Res 241:251–256
4. Segretain D, Rambourg A, Clermont Y (1981) Three dimensional arrangement of mitochondria and endoplasmic reticulum in the heart muscle fiber of the rat. Anat Rec 200(2):139–151
5. Denk W, Horstmann H (2004) Serial block-face scanning electron microscopy to reconstruct three-dimensional tissue nanostructure. PLoS Biol 2(11):e329
6. Heymann JA et al (2006) Site-specific 3D imaging of cells and tissues with a dual beam microscope. J Struct Biol 155(1):63–73
7. Dahl R et al (2015) Three-dimensional reconstruction of the human skeletal muscle mitochondrial network as a tool to assess mitochondrial content and structural organization. Acta Physiol (Oxf) 213(1):145–155
8. Glancy B et al (2015) Mitochondrial reticulum for cellular energy distribution in muscle. Nature 523(7562):617–620
9. Parry HA et al (2024) Impact of capillary and sarcolemmal proximity on mitochondrial structure and energetic function in skeletal muscle. J Physiol 602(9):1967–1986
10. Mekuč MŽ et al (2020) Automatic segmentation of mitochondria and endolysosomes in volumetric electron microscopy data. Comput Biol Med 119:103693
11. Conrad R, Narayan K (2023) Instance segmentation of mitochondria in electron microscopy images with a generalist deep learning model trained on a diverse dataset. Cell Systems 14(1):58–71.e55
13. Müller A, Schmidt D, Albrecht JP, Rieckert L, Otto M, Galicia Garcia LE, Fabig G, Solimena M, Weigert M (2024) Modular segmentation, spatial analysis and visualization of volume electron microscopy datasets. Nat Protoc 19:1436–1466. https://doi.org/10.1038/s41596-024-00957-5
12. Glancy B (2023) MitoNet: a generalizable model for segmentation of individual mitochondria within electron microscopy datasets. Cell Systems 14(1):7–8. https://doi.org/10.1016/j.cels.2022.12.004
14. Hayashi K, Miller RG, Brownell AKW (1989) Central core disease: ultrastructure of the sarcoplasmic reticulum and T-tubules. Muscle Nerve 12(2):95–102
15. Scarff CA, Fuller MJ, Thompson RF, Iadaza MG (2018) Variations on negative stain electron microscopy methods: tools for tackling challenging systems. J Vis Exp 132:57199
16. Kuchenbrod MT, Schubert US, Heintzmann R, Hoeppener S (2021) Revisiting staining of biological samples for electron microscopy: perspectives for recent research. Mater Horiz 8(3):685–699
17. Legland D, Arganda-Carreras I, Andrey P (2016) MorphoLibJ: integrated library and plugins for mathematical morphology with ImageJ. Bioinformatics 32(22):3532–3534
18. Püspöki Z, Storath M, Sage D, Unser M (2016) Transforms and operators for directional bioimage analysis: a survey. In: De Vos WH, Munck S, Timmermans JP (eds) Focus on Bio-Image Informatics. Springer, Cham Switzerland p 69–93
19. Domander R, Felder AA, Doube M (2021) BoneJ2-refactoring established research software. Wellcome Open Res 6:37
20. Sommer C, Straehle C, Köthe U, Hamprecht FA (2011) Ilastik: interactive learning and segmentation toolkit. In: 2011 IEEE International Symposium on Biomedical Imaging: From Nano to Macro; IEEE, Chicago IL p 230–233. https://doi.org/10.1109/ISBI.2011.5872394
21. Chung DJ et al (2022) Metabolic design in a mammalian model of extreme metabolism, the North American least shrew (Cryptotis parva). J Physiol 600(3):547–567

Chapter 15

Simultaneous Measurement of Mitophagic Flux and Mitochondrial Morphology During Myogenesis

Fasih A. Rahman and Joe Quadrilatero

Abstract

Mitophagy is the selective degradation of mitochondria via autophagy and is essential for maintaining mitochondrial quality and cellular homeostasis as well as being required for skeletal muscle differentiation and myogenesis. While traditional techniques such as immunoblotting, electron microscopy, and live-cell imaging provide valuable insights, they often offer only static or indirect measurements of mitophagy. Mitophagy flux assays capture dynamic turnover events and are, therefore, more informative. In this chapter, we present a streamlined method for monitoring and measuring mitophagic flux in C2C12 myoblasts during differentiation. Further, this approach allows for simultaneous quantification of mitochondrial morphology, providing a powerful platform for studying mitochondrial quality control in vitro.

Keywords Myogenesis, Mitophagic flux, Mitophagy, Mitochondrial quality control, Adenovirus, Fluorescence mitophagy reporter, Mitochondrial network analysis, Mitochondrial remodeling

1 Introduction

Skeletal muscle is a metabolically demanding tissue that undergoes extensive remodeling during development and regeneration. One of the most critical remodeling events is myogenesis, the tightly regulated process by which mononucleated skeletal muscle cells proliferate, differentiate, and fuse to form multinucleated myotubes (i.e., nascent skeletal muscle fibers) [1]. This transition involves dramatic shifts in cellular structure, gene expression, and metabolic state, particularly at the level of the mitochondria. During myogenesis, the selective degradation of damaged or redundant mitochondria via the autophagy machinery (termed mitophagy) plays a critical role in maintaining mitochondrial quality, promoting cellular homeostasis, and establishing a functionally competent and energetically efficient mitochondrial network necessary for mature skeletal muscle function [2].

Joe Quadrilatero (ed.), *Myogenesis and Skeletal Muscle Biology*, Methods in Molecular Biology, vol. 3057,
https://doi.org/10.1007/978-1-0716-5396-8_15,

As interest in mitophagy has grown, a wide array of techniques have been developed to investigate its regulation and biological relevance. Classical methods, such as electron microscopy (EM), have been used to identify mitophagosomes by morphology [3–5]. In addition, immunoblotting can reveal changes in mitophagy-related signaling pathways, while subcellular fractionation can help determine the localization of specific mitophagy regulators to mitochondria [4, 6, 7]. Live-cell imaging can also be performed for the visualization of mitochondrial trafficking and interactions with the autophagic machinery [8–10]. However, many of these methods offer only static or indirect insights and often require time-consuming sample preparation. More importantly, they fail to reflect the dynamic nature or rate of mitochondrial turnover under physiologically relevant conditions [4, 11]. In contrast, the measurement of mitophagy flux provides a more comprehensive and temporally resolved view of the process, encompassing key stages, such as mitochondrial tagging, autophagosome formation, lysosomal delivery, and degradation. Accurate quantification of mitophagy flux is, therefore, critical for understanding how cells adapt to stress, maintain energy balance, or deteriorate in disease. To this end, various in vitro approaches have been developed, with fluorescent mitophagy reporter systems, such as dual-tagged mitochondrial constructs, emerging as one of the most powerful and accessible tools [12–16]. These reporters can be introduced into cells using plasmid transfection or viral transduction systems, allowing either transient or stable expression depending on the experimental design. In the context of skeletal muscle, these systems provide a valuable means to track mitophagy progression throughout different stages of myogenic differentiation.

In this chapter, we describe a simplified, scalable method for generating, amplifying, and concentrating adenoviral constructs using low-speed centrifugation. This approach minimizes technical complexity and reduces processing time while maintaining high transduction efficiency in myoblasts. Moreover, it enables the simultaneous quantification of mitophagy flux and mitochondrial network parameters [15, 17], providing a robust platform to investigate the dynamic interplay between mitochondrial remodeling and quality control mechanisms in vitro.

2 Materials

1. C2C12 myoblast cells.
2. HEK293 cells.
3. Ad-COX8-eGFP-mCherry adenovirus.
4. 60 mm and 150 mm tissue culture plates.
5. 24-well tissue culture plates.

6. 15 mL and 50 mL sterile conical tubes.
7. Sterile cryovials.
8. 24 mm × 24 mm #1 coverslips.
9. Glass microscope slides.
10. Sterile pipette tips (various sizes).
11. Sterile serological pipettes (various sizes).
12. Dulbecco's Modified Eagle Medium with 1.0 g/L glucose, L-glutamine, and sodium pyruvate (DMEM).
13. Fetal bovine serum (FBS).
14. Serum plus II medium supplement (14009C, Sigma-Aldrich).
15. Horse serum.
16. Penicillin/Streptomycin (10,000 IU/mL penicillin, 10,000 μg/mL streptomycin).
17. 0.25% trypsin–EDTA.
18. 1× phosphate-buffered saline (PBS).
19. Cultrex basement membrane extract (3432-010-01; R&D Systems).
20. Hexadimethrine bromide (also known as polybrene).
21. Protamine sulfate.
22. Chondroitin sulfate C.
23. 37% formaldehyde.
24. 100% ethanol.
25. Fluoromount-G mounting medium or similar.
26. Objective immersion oil.
27. Tungsten steel engraving pen.
28. Rocking platform.
29. 37°C water bath.
30. Ice box.
31. -80°C freezer.
32. Centrifuge with appropriate rotors.
33. Cell culture incubator (maintained at 37°C, 5% CO_2).
34. Light microscope.
35. Fluorescent microscope or plate reader (e.g., Zeiss Axio Observer Z1, Agilent BioTek Cytation 5).
36. Confocal Microscope (e.g., Zeiss LSM 800).
37. Serum-free media: Add 5 mL Penicillin/Streptomycin to 500 mL of DMEM. Invert to mix and store at 4°C.
38. Trypsin working solution: Dilute 0.25% trypsin–EDTA 1:1 in PBS to obtain a 0.125% solution. Store at 4°C and use within 1 week.

39. 4% formaldehyde in PBS: Add 10.8 mL of 37% formaldehyde stock solution to 89.2 mL of PBS. Mix by inversion. The solution is stable at room temperature for several months (*see* **Note 1**).
40. Growth media: Add 28.09 mL Serum plus II, 28.09 mL FBS, and 5.62 mL Penicillin/Streptomycin to 500 mL of DMEM. Invert to mix and store at 4°C.
41. Differentiation media: Add 10.31 mL horse serum and 5.15 mL Penicillin/Streptomycin to 500 mL of DMEM. Invert to mix and store at 4°C.
42. Protamine sulfate stock solution (PS; 10 mg/mL): Add 100 mg protamine sulfate to 10 mL ultrapure water. Heat gently in a 37°C water bath with intermittent mixing until fully dissolved. Store at 4°C and use within 1 month.
43. Chondroitin sulfate C stock solution (CSC; 10 mg/mL): Add 100 mg chondroitin sulfate C to 10 mL ultrapure water. Heat gently in a 37°C water bath with intermittent mixing until fully dissolved. Store at 4°C and use within 1 month.
44. Hexadimethrine bromide stock solution (HB stock; 8 mg/mL): Add 80 mg hexadimethrine bromide to 10 mL ultrapure water. Optionally aliquot and store at −20°C or 4°C for 1 month.
45. Hexadimethrine bromide working solution (HB working; 8 μg/mL): Dilute HB stock 1:1000 in serum-free media (e.g., 10 μL of HB stock in 10 mL of serum-free media).
46. Coverslip preparation: Score and break 24 mm × 24 mm #1 coverslips into four equal squares using a tungsten steel engraving pen. Place coverslips in 100% ethanol in a sterile culture dish (e.g., 60 mm) and allow to air dry completely inside a biosafety cabinet. Coverslips can be used directly for cell plating after ethanol evaporation.

3 Methods

3.1 Cell Culture

1. Prepare 50 mL conical tubes with pre-warmed growth media and sterile PBS.
2. Warm all media to 37°C prior to use.
3. Rapidly thaw frozen cell aliquots (e.g., C2C12 or HEK293 cells) in a 37°C water bath.
4. Transfer thawed cells to a 15 mL conical tube containing 5 mL of prewarmed growth media.
5. Centrifuge cells at 500 × *g* for 3 min to pellet.
6. Carefully aspirate the supernatant and resuspend the cell pellet in 1–2 mL of fresh growth media.

7. For virus amplification, plate HEK293 cells on 150 mm tissue culture dishes containing 20 mL of growth media. Gently rock plates in a forward–backward and side-to-side motion to ensure even cell distribution.
8. For virus imaging and mitophagy flux experiments, plate cells (HEK293 or C2C12) on 24-well plates at 20,000/well.
9. Incubate cells at 37°C, 5% CO_2. Monitor daily for confluence and morphology using a light microscope.
10. For HEK293 cells, proceed with viral amplification at ~60% confluence. For C2C12 myoblasts, proceed with viral transduction at ~40% confluence (*see* **Note 2**).

3.2 Cultrex Coating of Coverslips

1. Using sterile technique, place cut coverslips (1/4 of a 24 mm × 24 mm #1 coverslip) into each well of a sterile 24-well tissue culture plate.
2. Rinse wells containing coverslips once with PBS.
3. Prepare a 1:200 dilution of Cultrex in cold sterile PBS (e.g., 50 μL Cultrex in 10 mL PBS for 24 wells). Ensure all solutions and tubes are kept on ice, as Cultrex polymerizes above 15°C.
4. Gently pipette the mixture inside the tube to mix. Do not vortex (*see* **Note 3**).
5. Add 400 μL of diluted Cultrex to each well, fully covering the coverslip.
6. Incubate plates at 37°C for at least 1 h to allow polymerization.
7. After incubation, remove the coating solution and wash wells once with cold PBS. Plates are now ready for cell seeding.

3.3 Viral Amplification

1. Thaw Ad-COX8-eGFP-mCherry viral stock aliquots at room temperature immediately before use.
2. Gently wash adherent HEK293 cells once with prewarmed sterile PBS.
3. Add 20 mL of fresh growth media to each 150 mm plate.
4. Add virus dropwise across the surface to ensure even distribution.
5. Incubate cells undisturbed at 37°C, 5% CO_2 for 2–5 days.
6. Monitor daily under a light microscope for cytopathic effect (CPE), characterized by cell rounding, detachment, and floating cells (*see* **Note 4**).

3.4 Viral Isolation and Purification

1. Collect both supernatant and cells (using Trypsin working solution) from virus-amplified HEK293 plates into 50 mL conical tubes.
2. Centrifuge at 1000 × *g* for 10 min to pellet the cells.
3. Transfer the supernatant to fresh 50 mL tubes and keep on ice.

4. Resuspend cell pellets in 2–5 mL of PBS.
5. Perform three to four freeze–thaw cycles by freezing the cells at −80°C for 10 min then thawing in a 37°C water bath for 10 min. Repeat.
6. After lysis, centrifuge the lysate at 1000 × *g* for 10 min to pellet debris.
7. Combine the cleared lysate with the previously collected supernatant.
8. Add protamine sulfate (PS) and chondroitin sulfate C (CSC) to final concentrations of 100 μg/mL each (*see* **Note 5**).
9. Incubate overnight at 4°C on a rocking platform to enhance viral precipitation.
10. The following day, centrifuge at 3000 × *g* for 45 min to pellet virus.
11. Carefully decant the supernatant, avoiding disturbance of the viral pellet.
12. Resuspend the pellet in 1/100th the original volume of PBS (e.g., 500 μL for 50 mL supernatant).
13. Aliquot and store virus at −80°C for long-term use.

3.5 Viral Titration and Transduction Optimization

1. Seed 20,000 C2C12 cells per well in a 24-well plate with coated coverslips.
2. The next day, thaw Ad-COX8-eGFP-mCherry virus aliquots at room temperature and mix thoroughly by pipetting.
3. Wash each well once with sterile PBS.
4. Add 200 μL of serum-free media containing 8 μg/mL hexadimethrine bromide (HB working solution) to each well.
5. Add 10 μL of serially diluted virus (e.g., 10^{-2} to 10^{-5}) to wells to test different multiplicities of infection (MOIs). Gently mix to ensure even distribution.
6. Incubate cells at 37°C, 5% CO_2 for 1–2 h, depending on MOI and cell type (*see* **Note 6**).
7. After incubation, aspirate viral media, wash cells once with PBS, and replace with 500 μL fresh growth media.
8. Allow cells to recover for 24 h. Transgene expression (e.g., fluorescent reporter) should be expressed at this point.
9. Image cells using a fluorescence microscope or plate reader (e.g., Agilent BioTek Cytation 5, Zeiss Axio Observer Z1).
10. Count the number of transduced cells per field.
11. Determine the number of imaging fields per well based on the well plate format used in the experiment (outlined in Table 1). Calculate viral titer (IFU/mL) using the formula:

Table 1
Derivation of area counted in fields/well (adapted from Adeno-X Rapid Titer Kit Protocol)

Digital magnification (10×)				Fields per well		
Objective lenses	Total magnification	Field diameter (mm)	Field area (mm^2)	12-well plate area = 3.8 cm^2	24-well plate area = 2.0 cm^2	96-well plate area = 0.32 cm^2
4×	40×	5	19.6	19	10	1.6
10×	100×	1.8	2.54	150	79	12.6
20×	200×	0.9	0.64	594	313	50

$$\textit{Infectious units} \quad (IFU/mL) = \frac{(\textit{infected cells per field})*(\textit{fields per well})}{(\textit{volume of virus})*(\textit{dilution factor})}.$$

3.6 Viral Transduction in Murine Myoblast Cells

1. Plate 20,000 C2C12 myoblasts per well in a 24-well plate containing Cultrex-coated coverslips.
2. The following day, thaw PS-CSC–precipitated Ad-COX8-eGFP-mCherry virus from the –80°C freezer at room temperature. Pipette gently to ensure homogeneity.
3. Wash the cells once with sterile PBS.
4. Add 200 μL of serum-free media containing 8.0 μg/mL hexadimethrine bromide (polybrene) to each well. Add virus at the desired MOI. Incubate at 37°C, 5% CO_2 for 1–2 h, depending on optimization. Gently rock the plate every 30 min to ensure even viral distribution.
5. After incubation, remove the viral media, wash cells once with PBS, and replace with 500 μL of fresh growth media.
6. The following day, cells should reach ~80%–90% confluence and begin expressing the fluorescent mitophagy reporter (i.e., Ad-COX8-eGFP-mCherry).
7. At this point, cells may be used directly, differentiated into myotubes (next section), or subjected to additional treatments as required for experimental endpoints (Fig. 1).

3.7 Differentiation of C2C12 Myoblasts

1. Once transduced C2C12 myoblasts reach ~80%–90% confluence in growth media, carefully aspirate the media and wash the cells twice with sterile PBS.

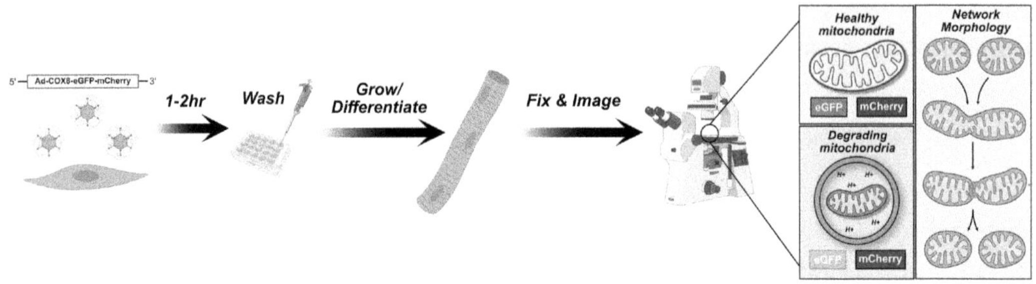

Fig. 1 Schematic overview of adenoviral transduction using Ad-COX8-eGFP-mCherry. Cells are incubated with Ad-COX8-eGFP-mCherry for 1–2 h, followed by PBS washes and replacement with fresh growth media. After transduction, cells can be subjected to treatment or differentiation, as illustrated, and subsequently fixed at defined time points for downstream image analysis

2. After the final wash, add differentiation media to each well.
3. Maintain cells in differentiation media at 37°C in a 5% CO_2 incubator.
4. Refresh differentiation media every 24 h.
5. Over the course of 2–5 days, monitor cells daily under a light microscope for evidence of myotube formation, indicated by elongated, multinucleated structures (*see* **Note 7**).
6. For mitophagy flux studies, cells may be fixed at desired time points (e.g., Day 1, Day 2, etc.) following differentiation to assess mitochondrial remodeling dynamics in newly formed myotubes.

3.8 Confocal Imaging

1. Wash transduced, adherent cells on coverslips twice with PBS.
2. Fix cells with 4% formaldehyde prepared in PBS for 10 min at room temperature.
3. Discard fixative and wash cells twice with PBS to remove residual formaldehyde.
4. Mount the coverslips onto standard glass microscope slides using Fluoromount-G mounting media.
5. Allow slides to dry overnight at room temperature in the dark to let the mounting medium harden.
6. Image cells the following day using a confocal microscope equipped with 488 nm and 561 nm lasers to excite eGFP and mCherry fluorophores, respectively.
7. Imaging was performed using a Zeiss LSM 800 confocal microscope with a 63× oil immersion objective.

3.9 Mitophagy Flux Analysis

1. Using ImageJ, open the merged fluorescence channel file (e.g., .tiff file).
2. Mitophagy flux was quantified using Color Threshold (Image > Adjust > Color Threshold) function to isolate red-only

(mCherry+/eGFP−) puncta, which indicate mitochondria localized to acidic lysosomes (Fig. 2).

3. Select the red channel, then set appropriate Hue/Saturation/Brightness parameters to highlight red-only puncta.
4. If normalizing per cell, press M to measure and get percent area occupied by red-only puncta.
5. If counting puncta is of interest, perform particle analysis (Analyze > Analyze Particles) to count puncta per cell or field.

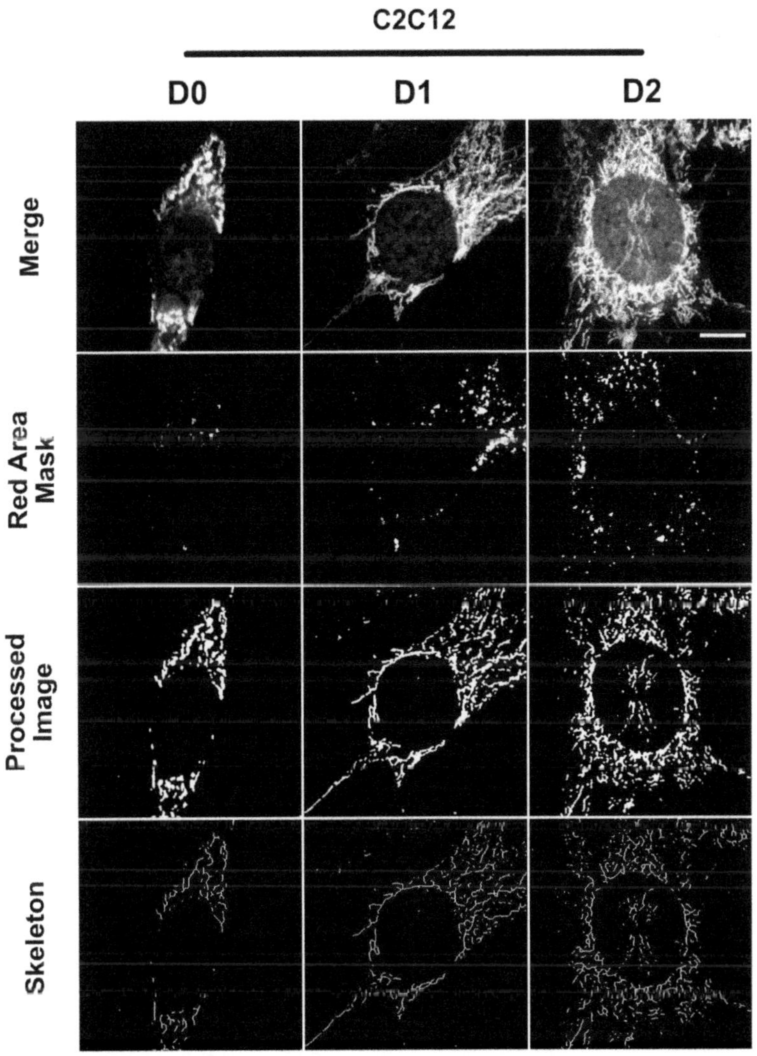

Fig. 2 Immunofluorescent images of proliferating (D0) and differentiating (D1 and D2) C2C12 myoblasts transduced with Ad-COX8-eGFP-mCherry. Green (eGFP) and red (mCherry) fluorescence mark the mitochondrial network, with red-only puncta indicating mitochondria undergoing degradation in acidic lysosomes. Overlap of green and red signals represents healthy non-degraded mitochondria. Images were analyzed using ImageJ. Representative red-only masks are shown as a measure of mitophagic flux. Processed and skeletonized images are also included to illustrate mitochondrial network structure. Scale bar = 10 μm

6. For normalization, calculate the red-only puncta as a proportion of total mitochondrial network footprint (i.e., total fluorescent area) to reflect mitophagy index.
7. At least five to ten fields per condition should be quantified.

3.10 Mitochondrial Network Analysis

1. Using ImageJ, open the merged fluorescence channel .tiff file.
2. Convert the image to 8-bit grayscale (Image > Type > 8-bit).
3. Run Iterative Deconvolution 3D plugin (Plugins > Deconvolution > Iterative Deconvolution 3D) using 5–15 iterations (*see* **Note 8**).
4. Adjust brightness and contrast uniformly across images to enhance mitochondrial signal (Image > Adjust > Brightness/Contrast).
5. Apply a Gaussian blur filter (Process > Filters > Gaussian Blur) with a sigma value of ~1.0–2.0 to reduce high-frequency noise.
6. Select the mitochondrial network using the threshold function (Image > Adjust > Threshold) (*see* **Note 9**).
7. Convert the processed image into a binary mask (Process > Binary > Make Binary).
8. Skeletonize the image using Skeletonize 2D/3D plugin (Plugins > Skeleton > Skeletonize 2D/3D; Fig. 2).
9. Run Analyze Skeleton (2D/3D) to extract data (*see* **Note 10**).

4 Notes

1. Formaldehyde is highly toxic. All handling should be performed inside a certified fume hood while wearing appropriate personal protective equipment (PPE), including a lab coat, gloves, and safety goggles.
2. HEK293 cells are more sensitive to viral load than C2C12 myoblasts. HEK293 growth slows significantly post-infection, therefore seeding a higher initial number of cells can help offset the impact of viral transduction.
3. Avoid vortexing Cultrex, as excessive agitation can cause premature polymerization. Gently invert the solution to mix and store on ice until use.
4. Fluorescence microscopy can be used to confirm viral expression (e.g., GFP or mCherry reporters) before harvest. Floating cells can also be transferred into a 96-well plate and screened for fluorescence using a plate reader.
5. If protamine sulfate or chondroitin sulfate C precipitate settles at the bottom of the tube, they can be resuspended by

gentle heating (e.g., 37°C) and vortexing. This has not been observed to negatively affect experimental outcomes.

6. Optimization of viral load and exposure time is necessary for each cell type. In our experience, removing the virus after 1–2 h improves cell viability and enhances expression of the fluorescent reporter.
7. Differentiation timing may vary depending on cell density and passage number. In our experience, C2C12 myoblasts exhibit optimal differentiation efficiency when used below passage 20 [18].
8. Increasing the number of deconvolution iterations can improve image sharpness but also increases processing time and may amplify noise. Be sure to use an appropriate point spread function (PSF) for the objective and fluorophore used.
9. Use consistent image processing parameters (e.g., thresholding, filters) across all samples to ensure reliable comparisons.
10. Skeleton analysis yields quantitative mitochondrial network data, including number of branches, average branch length, number of junctions, end-point voxels, and connectivity measures. These results are provided in both summary tables and annotated images and can be exported as .csv files for downstream statistical analysis.

Acknowledgments

Funding was provided to J.Q. by the Natural Science and Engineering Research Council of Canada (NSERC; RGPIN 258590). F.A.R was supported by an NSERC CGS-D scholarship.

References

1. Hawke TJ, Garry DJ (2001) Myogenic satellite cells: physiology to molecular biology. J Appl Physiol 91:534–551
2. Rahman FA, Quadrilatero J (2021) Emerging role of mitophagy in myoblast differentiation and skeletal muscle remodeling. Semin Cell Dev Biol 143:54–65
3. Chakraborty J, Caicci F, Roy M et al. (2020) Investigating mitochondrial autophagy by routine transmission electron microscopy: seeing is believing?. Pharmacol Res 160:105097
4. Klionsky DJ, Abdel-Aziz AK, Abdelfatah S et al. (2021) Guidelines for the use and interpretation of assays for monitoring autophagy (4th edition). Autophagy 17:1–382
5. Li Z, Wu Q, Liu L et al. (2021) Determination of mitophagy by electron microscope. Methods Cell Biol 165:103–110
6. Mizushima N, Yoshimori T (2007) How to interpret LC3 immunoblotting. Autophagy 3:542–545
7. Rahman FA, Graham MQ, Adam AM et al (2025) Mitophagy is required to protect against excessive skeletal muscle atrophy following hindlimb immobilization. J Biomed Sci 32(1):29
8. Karantanou C, Bibli S-I (2024) Monitoring mitophagy dynamics in live cells. Methods Mol Biol 2845:161–175
9. Moore AS, Holzbaur ELF (2019) Imaging the dynamics of mitophagy in live cells. Methods Mol Biol 1880:601–610

10. Wong YC, Holzbaur ELF (2015) Temporal dynamics of PARK2/parkin and OPTN/optineurin recruitment during the mitophagy of damaged mitochondria. Autophagy 11:422–424
11. Rahman FA, Baechler BL, Quadrilatero J (2024) Key considerations for investigating and interpreting autophagy in skeletal muscle. Autophagy 20(10):2121-2132
12. Allen GFG, Toth R, James J et al. (2013) Loss of iron triggers PINK1/Parkin-independent mitophagy. EMBO Rep 14:1127–1135
13. Kobayashi S, Patel J, Zhao F et al (2020) Novel dual-fluorescent mitophagy reporter reveals a reduced mitophagy flux in Type 1 diabetic mouse heart. J Am Osteopath Assoc 120:446–455
14. McWilliams TG, Prescott AR, Allen GFG et al. (2016) mito-QC illuminates mitophagy and mitochondrial architecture in vivo. J Cell Biol 214:333–345
15. Rahman FA, Friedrich Yap JM, Joseph TM et al. (2025) DNM1L-mediated fission governs mitophagy & mitochondrial biogenesis during myogenic differentiation. Cell Commun Signal 23: 158
16. Sun N, Malide D, Liu J et al. (2017) A fluorescence-based imaging method to measure in vitro and in vivo mitophagy using mt-Keima. Nat Protoc 12:1576–1587
17. Valente AJ, Maddalena LA, Robb EL et al. (2017) A simple ImageJ macro tool for analyzing mitochondrial network morphology in mammalian cell culture. Acta Histochem 119:315–326
18. Rahman FA, Hian-Cheong DJ, Boonstra K et al. (2024) Augmented mitochondrial apoptotic signaling impairs C2C12 myoblast differentiation following cellular aging through sequential passaging. J Cell Physiol 239:e31155

Check for updates

Chapter 16

Quantification of Mitochondrial Oxidative Phosphorylation in Skeletal Muscle

Aditya N. Brahmbhatt and Christopher G. R. Perry

Abstract

Mitochondrial oxygen consumption stimulated by ADP and inorganic phosphate is often measured as an index of ATP synthesis derived from oxidative phosphorylation. This measurement is typically performed by monitoring the rate at which oxygen disappears from a solution containing isolated mitochondria or either intact or permeabilized cells and tissue. Termed "respirometry," this method has become increasingly popular in numerous research disciplines focused on cellular metabolism, including the broader field of skeletal muscle physiology. Here, we describe how respirometry is performed in permeabilized skeletal muscle fiber bundles. This technique involves mechanically separating skeletal muscle fibers followed by exposure to a cell membrane-permeabilizing agent (e.g., saponin). This technique is thought to preserve the mitochondrial network, morphology, and interactions with surrounding intracellular structures. The method can be performed on small tissue samples and permits flexibility in designing substrate titration protocols to target specific mitochondrial pathways.

Keywords Respirometry, Permeabilized fibers bundles, Skeletal muscle, Mitochondria, Polarographic oxygen sensor

1 Introduction

Mitochondria are critical regulators of ATP synthesis in skeletal muscle. While measurements of ATP content in cell or tissue samples are often performed with spectrophotometric, spectrofluorometric, or bioluminescent technologies, the direct measurement of ATP synthesis in real-time using mitochondrial preparations has historically been limited by a lack of kinetic methodologies. Rather, assessments of oxygen consumption have long been used as an index of mitochondrial oxidative phosphorylation. The principle depends on the design of substrate titration protocols in mitochondrial preparations, and the interpretation of the results is based on the chemiosmotic theory of how mitochondrial oxidative phosphorylation is regulated.

For example, electrons harnessed from substrates, including glucose, fatty acids, and certain amino acids, are shuttled primarily

Joe Quadrilatero (ed.), *Myogenesis and Skeletal Muscle Biology*, Methods in Molecular Biology, vol. 3057,
https://doi.org/10.1007/978-1-0716-5396-8_16,

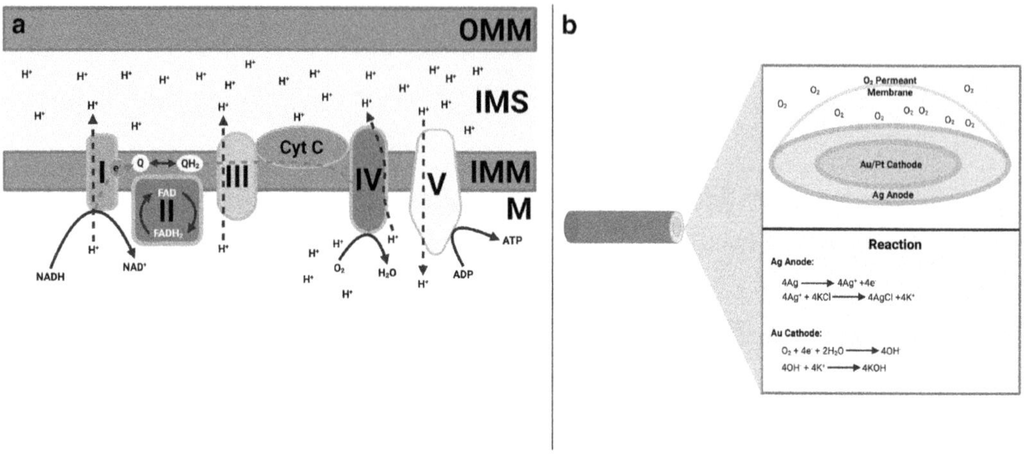

Fig. 1 Chemiosmotic theory. (**a**) Reducing equivalents in the form of nicotinamide adenine dinucleotide (NADH) and flavin adenine dinucleotide ($FADH_2$) donate electrons to Complexes I and II (I and II), respectively. Not shown: $FADH_2$ from beta-oxidation donates electrons to the electron-transferring flavoprotein. Two mobile electron carriers, ubiquinone (Q) and cytochrome *c* (cyt C), help to shuttle electrons in between the major Complexes. The transfer of these electrons across the four Complexes facilitates the pumping of H^+ from the mitochondrial matrix (M) to the intermembrane space (IMS), generating an H^+ gradient. Finally, the electrons are transferred to oxygen, which combines with hydrogen to form H_2O at Complex IV (IV). Complex V (V), also known as ATP synthase, couples the energy inherent in proton re-entry to the matrix to phosphorylate ADP to ATP. (**b**) Polarographic oxygen sensor design and oxygen detection mechanism. As the silver anode donates its electrons to the oxygen molecules at the gold cathode, a current is generated that is directed to the computer hardware. When the oxygen concentration changes, the amount of current changes proportionally. The rate of change in current is used by the software to calculate "oxygen flux." A negative change in oxygen concentration in the chamber is interpreted as mitochondrial oxygen consumption. Complex III (III); inner mitochondrial membrane (IMM); outer mitochondrial membrane (OMM). Made using BioRender

to Complexes I (NADH), II ($FADH_2$ from the tricarboxylic acid (TCA) cycle), and the electron-transferring flavoprotein ($FADH_2$ from beta oxidation of fatty acids). As predicted from the chemiosmotic theory, the flux of electrons to oxygen at Complex IV promotes proton pumping from the matrix to intermembrane space through Complexes I, III, and IV. As protons accumulate in this space (Fig. 1a), they contribute to the proton motive force that drives proton re-entry through ATP synthase when activated by ADP and inorganic phosphate (Pi). As both ADP and Pi rise during periods of heightened ATP hydrolysis from ATP-consuming proteins, such as during skeletal muscle contraction, this system represents a feedback loop that links energy demand represented by ATP turnover to ATP synthesis. The key mediators in this process include the ratio of (ADP, Pi)/(ATP), NAD+/NADH, and FAD/$FADH_2$, which determine the state of activity in key dehydrogenases that transfer electrons from substrates to these reducing equivalents as well as ATP synthase itself. In this way, ADP-stimulated oxygen consumption or "respiration" is an index of oxidative phosphorylation of ADP to

ATP and is measured through a technique called "respirometry" by monitoring the disappearance of oxygen from a solution containing mitochondrial preparations.

Plate-based technologies using oxygen-sensitive fluorophores and fluorometric detectors are appropriate for intact adherent cells and have been used extensively for isolated mitochondria. Suspension-based respirometers are often used for isolated mitochondria as well as permeabilized or intact cells and tissues. In suspension respirometry, the sample is suspended in a buffer that is constantly stirred to ensure uniform distribution of oxygen across the closed chamber. A polarographic oxygen sensor within the chamber often consists of a polarized gold cathode and a silver anode in a potassium chloride solution sealed with an oxygen-permeant membrane [1]. As oxygen diffuses through the membrane, it reacts with the gold or platinum cathode in conjunction with a series of reactions facilitated by a silver anode, to produce electrons (Fig. 1b). This process produces an electrical current flowing from the sensor to a computer with software that equates current to oxygen concentration in the solution. The electrical signal generated at the electrode is proportional to the amount of oxygen such that the rate of oxygen disappearance from solution can be quantified and interpreted as oxygen consumption. The use of substrates that stimulate mitochondrial oxygen consumption, such as those derived from glucose, fatty acids, or amino acids, guides the interpretation that changes in oxygen disappearance are due solely to mitochondrial oxygen consumption. The use of very small tissue samples, as is often the case with permeabilized skeletal muscle fibers, is made possible by high-resolution respirometers such as the Oroboros Oxygraph-2K. This system contains a tightly sealed system that provides a stable signal with low background noise. Such systems can detect oxygen consumption rates as low as 0.001 $\mu M \cdot s^{-1}$ [1].

There are four common methods of sample preparation that can be used for respirometry—isolated mitochondria, permeabilized fiber bundles (PmFBs), tissue homogenate, and intact cells [2]. Each method provides varying levels of physiological relevance, which is important to consider when interpreting the results. Ultimately, the method of preparation an experimenter chooses depends on the experimental objective (see Table 1). This chapter will focus on permeabilized skeletal muscle fiber bundles.

2 Materials

In order to model the *in-vivo* ionic environment, the pH of all buffers and substrates should be maintained close to physiological pH (*see* **Note 2**). In addition, all buffers and substrates can be stored at −20°C for at least 6 months. The design of buffers outlined below is

Table 1
Comparison of different mitochondrial preparation techniques used to assess mitochondrial respiration

Isolated mitochondria	Permeabilized fiber bundles	Homogenate	Intact cells
Advantages: • Can assess function from two distinct populations of mitochondria (e.g., subsarcolemmal vs intermyofibrillar) [19]. • Reduces variability across replicates compared to PmFB by providing a more homogeneous mixture of mitochondria within the tissue sample [3]. • Minimal oxygen diffusion limitations to electron transport chain [3, 19]. • Removes plasma membrane barrier allowing exogenous substrates to enter the mitochondria [2]. **Disadvantages:** • Requires high amounts of tissue (≥100–150 mg,) [3, 19]. • Care is required as the preparation may damage mitochondrial structure (outer membrane) and network, which could alter function [2, 19, 37]. • Loses physiological tractability due to removal of other cytosolic structures (e.g., sarcoplasmic reticulum and myofibrils) which the mitochondria interact with in-vivo [2, 38, 39]. • Isolation procedure may not obtain all mitochondria [38].	**Advantages:** • Better preserves mitochondrial network, morphology and interactions between mitochondria and cytoskeletal structures which may influence mitochondrial respiratory function [3, 5, 19]. • Wash step removes influence of pre-existing cytoplasmic substrates and adenylates [19, 40]. • Requires less tissue (~1.5–4 mg) compared to isolated mitochondria [3, 19]. • Permeabilizes plasma membrane barrier, allowing exogenous substrates to enter the mitochondria [2]. **Disadvantages:** • Depending on muscle, fiber-type heterogeneity between two bundles from the same muscle could increase variability between replicates [3, 41] • Requires high concentrations of oxygen to overcome oxygen diffusion limitations (*see* **Note 1**) [42]. • Technical expertise of experimenter in sample preparation may affect the outer mitochondrial membrane integrity and thus affect results.	**Advantages:** • Reduces variability (as long as the degree of homogenization is the same across samples) across replicates compared to PmFB by providing a more homogeneous mixture of mitochondria within the tissue sample [41, 43]. • Oxygen diffusion to mitochondria may not be as limiting compared to PmFB [43]. • Disintegrates the plasma membrane, allowing exogenously added substrates to enter the mitochondria. **Disadvantages:** • It is not possible to remove the influence of pre-existing endogenous substrates and adenylates may have on mitochondrial respiration. • Preparation may damage mitochondrial structure (outer membrane) and network, which can alter function [41].	**Advantages:** • Allows for assessment of mitochondrial function in its native cellular environment in the presence of important upstream metabolic enzymes in the cytosol (e.g., glycolytic enzymes), sarcolemma transporters of substrates, and physiological concentrations of endogenous substrates [1, 2]. **Disadvantages:** • Cannot assess maximal capacity as ADP and some substrates cannot cross the plasma membrane [2, 44]. • Does not control for influence of pre-existing endogenous substrates and adenylates may have on mitochondrial respiration. • Respiration is likely underestimated in intact tissue samples given the oxygen will not diffuse to the core of the tissue sample unless it is very thin

summarized from work by Erich Gnaiger's group [1, 3, 4]. Briefly, BIOPS is designed to model the extracellular ionic, osmotic, and pH environment while Buffer Z or MiRO model the intracellular environment. As such, BIOPS is used when separating skeletal muscle fibers with forceps and during permeabilization with saponin, whereas Buffer Z or MiRO are used for washing saponin from fibers and performing the respirometry procedures after the membrane has been permeabilized.

2.1 Skeletal Muscle Fiber Separation

1. 2× fine jeweller angular forceps.
2. Scalpel (recommend straight edge).

2.2 Permeabilization

1. Stock 100 mM CaK_2EGTA: Dissolve 2.002 g of $CaCO_3$ in 100 mM hot (80°C) solution of EGTA (7.608 g of EGTA in 200 mL ddH_2O). Add 2.3 g of KOH and add more as needed to reach a pH of 7.0.
2. Stock 100 mM K_2EGTA: Dissolve 7.608 g EGTA and 2.3 g KOH into 200 mL ddH_2O. Add more KOH as needed to reach a pH of 7.0.
3. BIOPS: Add 55.4 mL of stock 100 mM CaK_2EGTA (2.77 mM) and 144.6 mL of stock 100 mM K_2EGTA (7.23 mM) to 1500 mL of ddH_2O. Weigh out 6.99 g of Na_2ATP (5.77 mM), 2.67 g of $MgCl_2$•$6H_2O$ (6.56 mM), 5.02 g of taurine (20 mM), 9.81 g of Na_2Phosphocreatine (15 mM), 2.72 g of imidazole (20 mM), 0.154 g of dithiothreitol (0.5 mM), 19.52 g of MES hydrate (50 mM). Add aforementioned powder chemicals and once dissolved, adjust pH to 7.2 using KOH pellets. Bring up volume to 2000 mL with ddH_2O. Filter and aliquot into 50 mL tubes.
4. Saponin: prepare a stock concentration of 10 mg/mL

2.3 Wash and Mitochondrial Respiration Buffer

Of the several buffers used to sustain mitochondrial and tissue preparations during respirometry [11], the commonly used formulations of Buffer Z and MiRO are described below. These buffers are intended to model the intracellular environment's higher potassium concentration, lower sodium concentrations, osmolarity, low calcium, and pH compared to the extracellular environment. The buffers are used to wash the fibers following permeabilization as well as to support the tissue when performing experiments in a respirometer.

1. Buffer Z: Weigh 12.26 g of K-MES (105 mM), 1.12 g of KCl (30 mM), 0.7 g KH_2PO_4 (10 mM), 0.51 g of $MgCl_2$•$6H_2O$ (5 mM), 0.19 g of EGTA (1 mM), and 2.5 g of BSA (5 mg/mL; ideally fatty acid free) and add to 400 mL of ddH_2O. Adjust pH to 7.2 using KOH pellets, then bring volume to 500 mL with ddH_2O. Filter using 0.22 micron filters (vacuum is recommended) and aliquot into 50 mL tubes.

2. MiRO: Add 240 mL of K-Lactobionate (60 mM; see below) to 1500 mL of ddH_2O. Weigh out 0.38 g of EGTA (0.5 mM), 1.22 g of $MgCl_2 \bullet 6H_2O$ (3 mM), 5.02 g of taurine (20 mM), 2.72 g of KH_2PO_4 (10 mM), 9.54 g of HEPES (20 mM), 75.3 g of sucrose (110 mM), 2 g of BSA (1 g/L). Add aforementioned powder chemicals, and once dissolved, adjust pH to 7.2 using KOH pellets. Bring up volume to 2000 mL with ddH_2O. Filter using 0.22 micron filters and aliquot into 50 mL tubes.
3. Stock K-lactobionate (only for MiRO): Dissolve 71.6 g lactobionic acid in 300 mL ddH_2O. Adjust pH to 7.0 and bring up final volume to 400 mL with ddH_2O.

2.4 Substrates, Inhibitors, and Uncouplers

See Table 2 and Fig. 2.

3 Methods

Mitochondrial structure is an important determinant of its function [5]. Therefore, mitochondrial structural integrity must be preserved throughout the process of tissue preparation. This is achieved through gentle separation of fibers with sufficient practice to minimize time, and to keep the sample conditions cold (except when placed in the respirometer chamber) to prevent mechanical damage and degradation of the contents within the tissue. A fully trained and experienced experimenter could aim for an average of 2.5 min to prepare one bundle without sacrificing respiratory functional integrity.

3.1 Respirometer Set-Up and Polarographic Oxygen Sensor Calibration

This section refers to the use of the Oroboros O2k High-Resolution Respirometer (www.oroboros.at)

1. Calibrate polarographic oxygen sensor following manufacturer's instructions (*see* **Note 8**).
2. Following calibration, follow the manufacturer's instructions to set up the respirometer for an experiment. During this process, the experimenter has control over the chamber contents and conditions, such as choice of intracellular/respiration buffer (MiRO5 or Buffer Z) (*see* **Note 9**), temperature (*see* **Note 10**), and presence/absence of creatine and phosphocreatine (*see* **Note 11**).

3.2 Tissue Harvest

3.2.1 Rodent

1. Anesthetize the animal using 4%–5% vaporized isoflurane (*see* **Note 12**).
2. Skin the limb of the animal in which the skeletal muscle of interest is located and carefully isolate and dissect out the muscle.

Table 2
List of commonly used chemicals and substrates used in respirometry

Substrate	Site of action and purpose	Stock concentration(s) and conditions	Typical concentration(s) in the respirometer
Blebbistatin	• **Site of Action:** Myosin II ATPase • **Purpose:** Myosin Inhibitor (*see* **Note 3**)	10 mM; made in DMSO, no pH adjustment necessary, and store at −20°C in small aliquots to avoid more than one freeze–thaw cycle	5 μM
Pyruvate	• **Site of Action:** Pyruvate Dehydrogenase (PDH) • **Purpose:** Generates NADH and acetyl-CoA, which enters the TCA cycle (Fig. 2a).	2 M; make fresh every day in dH_2O	5 mM
Malate	• **Site of Action:** Malate Dehydrogenase • **Purpose:** Generates NADH and oxaloacetate, which when combined with Acetyl-CoA starts TCA cycle (Fig. 2a).	1 M; made in dH_2O, pH to 7, and store at −20°C in small aliquots to avoid more than one freeze–thaw cycle	2 mM
Glutamate	• **Site of Action:** Glutamate Dehydrogenase • **Purpose:** Generates NADH and α-ketoglutarate which, when combined with Acetyl-CoA starts TCA cycle (Fig. 2a).	2 M; made in dH_2O, pH to 7, and store at −20°C in small aliquots to avoid more than one freeze–thaw cycle	10 mM
Succinate	• **Site of Action:** Succinate Dehydrogenase • **Purpose:** Generates $FADH_2$ and fumarate, which is a TCA cycle intermediate (Fig. 2a).	2 M; made in dH_2O, pH to 7, and store at −20°C in small aliquots to avoid more than one freeze–thaw cycle	20 mM
Palmitoyl CoA (PCoA)	• **Site of Action:** ß-oxidation • **Purpose:** Generates NADH and $FADH_2$, as well as Acetyl-CoA, which enters the TCA cycle (Fig. 2b).	10 mM; made in dH_2O in original supplier's bottle, no pH adjustment necessary, and store at −20°C in small aliquots to avoid more than one freeze–thaw cycle	5–150 μM

(continued)

Table 2
continued

Substrate	Site of action and purpose	Stock concentration(s) and conditions	Typical concentration(s) in the respirometer
L-Carnitine	• **Site of Action:** Carnitine palmitoyltransferase I and II • **Purpose:** Supports transport of Palmitoyl CoA across the inner and outer mitochondrial membranes via the carnitinepalmitoyl transferase (CPT) system (Fig. 2b).	10 mM; made in dH_2O, pH to 7, and store at −20°C in small aliquots to avoid more than one freeze–thaw cycle	750–2000 μM
Malonyl CoA	• **Site of Action:** Carnitine palmitoyltransferase I (CPT I) • **Purpose:** Inhibitor (Fig. 2b) (*see* **Note 4**).	1 mM; made in dH_2O, most stable at −20°C at pH 5; store at −20°C in small aliquots to avoid more than one freeze–thaw cycle	0.5–20 μM
Palmitoylcarnitine	• **Site of Action:** ß-oxidation • **Purpose:** Bypasses CPT I-mediated transport across the outer mitochondrial membrane. Generates NADH and $FADH_2$ as well as Acetyl-CoA, which enters the TCA cycle (Fig. 2b).	10 mM; made in dH_2O in original supplier's bottle, no pH adjustment necessary, and store at −20°C in small aliquots to avoid more than one freeze–thaw cycle	10–80 μM
ADP	• **Site of Action:** Complex V/ATP Synthase • **Purpose:** Generates ATP	5 mM, 50 mM, and 500 mM; made in dH_2O, pH to 7, and store at −20°C in small aliquots to avoid more than one freeze–thaw cycle	25–7000 μM
Cytochrome *c*	• **Site of Action:** Inner mitochondrial membrane • **Purpose:** outer mitochondrial membrane integrity test (*see* **Note 5**).	4 mM, made in dH_2O, no pH adjustment necessary, and store at −20°C in small aliquots to avoid more than one freeze–thaw cycle	10 μM

(continued)

Table 2
continued

Substrate	Site of action and purpose	Stock concentration(s) and conditions	Typical concentration(s) in the respirometer
Carbonyl cyanide-p-trifluoromethoxy-phenylhydrazone (FCCP)	• **Site of Action:** Inner mitochondrial membrane • **Purpose:** Uncouples electron flux across the ETC from ATP production, allowing for the assessment of maximal ETC flux (*see* **Note 6**).	1 mM, made in DMSO, no pH adjustment necessary, and store at −20°C in small aliquots to avoid more than one freeze–thaw cycle	0.5–1 μM
Rotenone	• **Site of Action:** Complex I • **Purpose:** Inhibitor (*see* **Note** 7)	1 mM, made in ethanol abs., no pH adjustment necessary, and store at −20°C in small aliquots to avoid more than one freeze–thaw cycle	0.5 μM
Antimycin A	• **Site of Action:** Complex III • **Purpose:** Inhibitor (*see* **Note** 7)	5 mM, made in ethanol abs., no pH adjustment necessary, and store at −20°C in small aliquots to avoid more than one freeze–thaw cycle	2.5 μM
Oligomycin	• **Site of Action:** Complex V/ATP Synthase • **Purpose:** Inhibitor (*see* **Note** 7)	5 mM stock made in original supplier's bottle in ethanol abs. from which a further 0.01 mM stock solution can be made in ethanol abs., no pH adjustment necessary, and store at −20°C in small aliquots to avoid more than one freeze–thaw cycle	5–10 nM

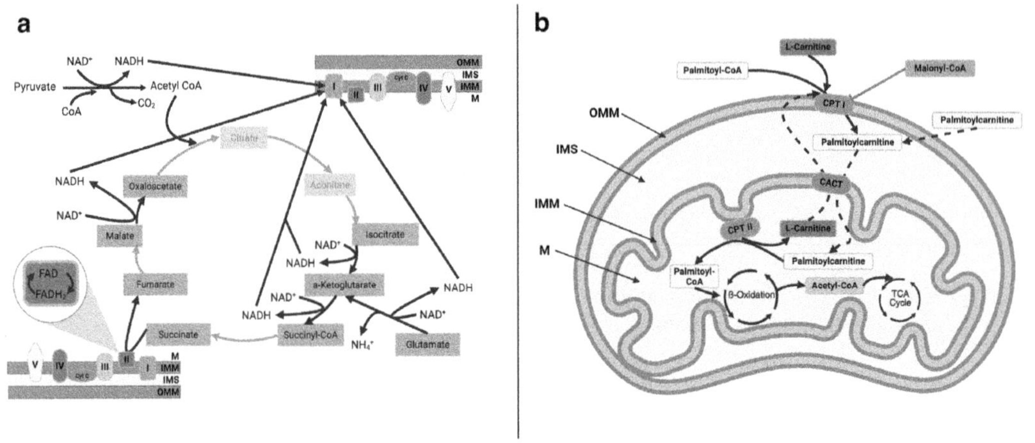

Fig. 2 Influence of substrates on energy pathways in the mitochondria. (**a**) Pyruvate generates acetyl-CoA which combines with oxaloacetate to form citrate in the TCA cycle. Malate provides a continuous supply of oxaloacetate to facilitate sustained pyruvate oxidation by preventing an accumulation of acetyl-CoA. NADH is produced during this process, which then donates its electrons to Complex I. While $FADH_2$ is produced within Complex II itself during the conversion of succinate to fumarate, passing along its electrons to coenzyme Q. (**b**) Palmitoyl-CoA must first combine with L-carnitine via carnitine palmitoyl transferase I (CPT I) to cross the outer mitochondrial membrane (OMM). It then crosses the inner mitochondrial membrane (IMM) via carnitine acylcarnitine transferase (CACT) in exchange for L-carnitine going in the opposite direction. Inside the matrix (M), it is converted back to palmitoyl-CoA via carnitinepalmitoyl transferase II (CPT II) bound to the inner leaflet of the inner mitochondrial membrane. Endogenous malonyl-CoA can be produced during high TCA cycle flux and can inhibit CPT I; therefore, exogenous administration of this inhibitor can be used to assess CPT I sensitivity to inhibition by malonyl-CoA. When carnitine is pre-conjugated to palmitoyl-CoA in the form of palmitoylcarnitine from commercial suppliers, its addition to the respiration buffer bypasses CPT I. This test provides insight into downstream pathways given palmitoyl carnitine will subsequently be converted to palmitoyl-CoA by CPT II prior to entering ß-oxidation to produce acetyl-CoA, NADH, and $FADH_2$. Complex I (I); complex II (II); complex III (III); complex IV (IV); complex V/ATP synthase (V); intermembrane space (IMS). Made using BioRender

3. Immediately place skeletal muscle in a cell culture dish (150×25 mm) containing BIOPS buffer on ice following dissection.

3.2.2 Human

1. Apply local anesthetic (2% lidocaine, 0.005 mg/mL epinephrine) subcutaneously (*see* **Note 13**).
2. Obtain skeletal muscle biopsy using a percutaneous needle from the muscle of interest (*see* **Note 14**).
3. Immediately place muscle in a cell culture dish (150×25 mm) containing BIOPS buffer on ice following biopsy.

3.3 Fiber Separation

1. Place the cell culture dish with skeletal muscle tissue in BIOPS buffer under a 1× microscope with a light source on a flat, frozen metal block or ice-pack (Fig. 3a and b) (*see* **Note 15**).
2. Carefully cut down the muscle into smaller strips with the scalpel using a gentle "see-saw" or "press and wiggle" motion (*see*

Fig. 3 Bundle fiber separating station set-up and wet bundle weighing. (**a**) 1× microscope with light source. (**b**) Dark background placed under separating dish providing the best visibility of the fiber bundles while separating under the microscope. (**c, d**) 1.5 mL Eppendorf tube with cold BIOPS buffer filled to the very top to create a convex meniscus held by surface tension (**c**), upon which the fiber bundle is placed to obtain wet weight (**d**). (**e**) Dabbing off excess BIOPS fluid from fiber bundle on delicate tissue to prevent overestimation of wet bundle weight

Note 16). It is critical to avoid stretching fibers along their longitudinal axis (avoid "pulling") as this could tear the mitochondrial network.

3. Using the fine angular forceps, gently begin separating the fibers, starting from the center and working all around the bundle (*see* **Note 17**). Continue separating until you see adequate spacing in-between fibers (*see* **Note 18**).
4. Place bundle in its own individual tube filled with cold BIOPS.

3.4 Wet Bundle Weighing

Weighing the bundles using this method can be used to help normalize rates and also to help judge the best size to get the cleanest possible trace and best rates (*see* **Note 19**).

1. Put a 1.5 mL Eppendorf tube in a tube holder to keep upright and fill with cold BIOPS buffer to the very top of the tube till it forms a dome held together by surface tension (Fig. 3c). Place the tube on a weigh scale and "zero" the scale.
2. Pick up skeletal muscle fiber bundle using the fine angular forceps and dab off excess fluid on a lint-free delicate tissue wipe (such as Kimwipe©) (Fig. 3d) (*see* **Note 20**).
3. Gently leave bundle on top of the convex meniscus of the BIOPS buffer, which forms when the tube is completely filled

with buffer. Record weight once the value displayed on the balance is stabilized (Fig. 3e).

4. If the bundle size is too big (*see* **Note 19**), gently remove some additional fibers as needed and repeat using Steps 1–3.

3.5 Permeabilization and Wash

1. Prepare 10 mg/mL saponin solution (*see* **Note 21**).
2. Fill 1.5 mL Eppendorf tube (one for each sample) with 1.5 mL of cold BIOPS buffer and add fresh saponin to BIOPS at a concentration of 40 μg/mL (6 μL of 10 mg/mL saponin) for rodent skeletal muscle or 30 μg/mL (4.5 μL of 10 mg/mL saponin) for human skeletal muscle.
3. Transfer bundles into their respective permeabilization tubes and leave on a nutator at 4°C for 30 min (*see* **Note 22**).
4. Prepare 1.5 mL Eppendorf tubes with 1.5 mL of wash buffer (Buffer Z or MiRO).
5. Transfer bundles into their respective wash tubes and leave on a nutator at 4°C for 15 min (*see* **Note 22**).

3.6 Conducting the Respirometry Protocol

1. Before starting protocol, add 5 μM blebbistatin to the chamber (*see* **Note 3**).
2. Add bundle to chamber.
3. Add 100% oxygen into chamber to ensure adequate initial oxygen levels to avoid reoxygenation (*see* **Note 1**).
4. Allow approximately 10–15 min to let the oxygen concentration signal and flux stabilize.
5. Record background respiration of the bundle.
6. Begin protocol (*see* **Note 23**) by adding the first substrate into the chamber, wait for 2–3 min to let the oxygen flux stabilize, then record. If the manufacturer's software allows, create a mark at the time point that the substrate was added.
7. Repeat Step 5 and continue the protocol.
8. If normalizing data by dry bundle weight, make sure to recover and lyophilize bundle for at least 4 h after completion of protocol. A high-sensitivity balance with resolution to 0.001 mg should be used for dry weights if this method is chosen as an alternative to normalizing to wet weights.
9. Clean the chamber by removing the intracellular/respiration buffer, then performing three washes with ddH_2O, then three washes with 70% ethanol solution, and three washes again with ddH_2O.

3.7 Analysis

1. Record oxygen flux or calculate it from oxygen consumption data (*see* **Note 24**).

2. Normalize oxygen flux data using method of choice (*see* **Note 25**).

4 Notes

1. Two main factors to consider regarding starting oxygen concentrations within the chamber include oxygen diffusion into PmFB and the duration of respirometry protocol. The presence of intracellular structures including the cytoskeleton is thought to limit the rate at which oxygen diffuses from the respirometric buffer to the mitochondria. Adequate oxygen concentrations are needed to overcome this limitation. By generating a large enough pressure gradient across the sarcolemma, this ensures adequate diffusion into the cytoplasmic compartment and mitochondria. The lower limit of effective oxygen concentration may vary depending on the respirometer. It can be estimated by leaving a test PmFB sample in the chamber with substrates and ADP and noting the chamber oxygen concentration at which oxygen flux begins to decline. Furthermore, adequate oxygenation is also needed to ensure that a sufficiently high partial pressure of oxygen (P_{O_2}) is sustained throughout the length of the protocol (typically 45–60 min). Ideally, the chamber could be reopened and reoxygenated, and the flux would return to previous levels. However, anecdotally, the oxygen consumption rate following reoxygenation part way through the protocol does not always equal the rate observed prior to reoxygenation (Fig. 4). In our experience, reoxygenation should be avoided as much as possible by ensuring a high oxygen concentration at the beginning of the procedure, optimizing bundle weights to as small as possible without sacrificing reproducibility or sensitivity of measurements, and/or limiting the duration of the titration protocol.
2. The pH of BIOPS and MiRO or Buffer Z buffers have been designed to model the extracellular and intracellular environments, respectively. Using proton nuclear magnetic resonance and chemical exchange saturation transfer imaging, intracellular pH was estimated to be 6.97–7.10 in human skeletal muscle [6] and 7.25 in mouse skeletal muscle [7], respectively. Microdialysis assessment of interstitial (extracellular) pH in human skeletal muscle has been estimated to be 7.17–7.38 [8, 9]. Given this range of pH values between the two compartments, it is recommended to maintain the pH of buffers around 7.2. Accordingly, most substrates should also have a pH of ~7 to avoid large fluctuations in the intracellular/

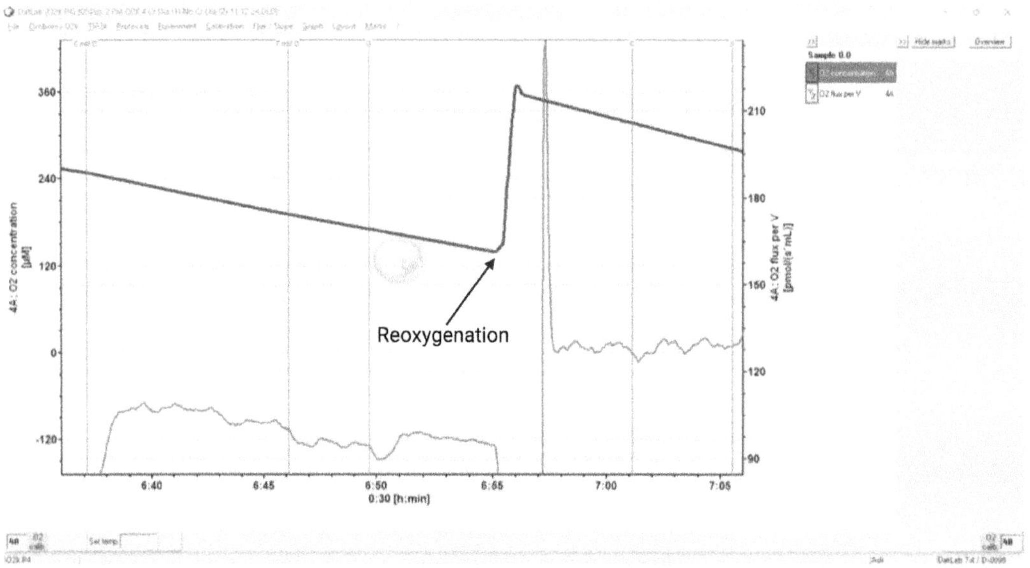

Fig. 4 Representative trace from DatLab 7.4 software used for Oroboros Oxygraph-2K. Both oxygen concentration (blue line) and oxygen flux (red line) are shown. While displaying oxygen concentration, this software also calculates and displays oxygen flux simultaneously. After reoxygenation of the chamber (indicated by black arrow), oxygen flux may not return to its previous level

respiration buffer pH in the respirometer. Anecdotally, this is particularly important for promoting solubility when preparing stock ADP. Likewise, unadjusted acidic metabolites, such as ADP, malate, and succinate, may greatly impact the pH of the respiration buffers.

3. ADP causes permeabilized skeletal muscle fibers to contract by binding myosin. This contraction increases respiration at lower ADP concentrations, thereby reflecting a contraction-induced increase in mitochondrial ADP sensitivity, but may lower maximal respiration in certain muscle types and experimental conditions that could be due to limited exposure of each fiber to the respiration buffer [10]. This effect is important to consider within the context of the experimental question in a given study. For example, if mitochondrial oxygen consumption is being questioned in the context of relaxed skeletal muscle, this phenomenon can be prevented by including a myosin II ATPase inhibitor known as "blebbistatin". To date, no myosin I ATPase inhibitor has been validated in permeabilized fibers. In the presence of blebbistatin, PmFBs remain partially relaxed, which may create conditions that are more reflective of the in vivo state of relaxed skeletal muscle.

4. Malonyl-CoA can be produced endogenously from acetyl-CoA by the enzyme acetyl-CoA carboxylase. An accumulation

of malonyl-CoA can inhibit CPT I-mediated transport of palmitoyl-CoA (PCoA) across the outer mitochondrial membrane. Titration of malonyl-CoA will lower PCoA-stimulated respiration and can provide an indication of sensitivity of CPT I to its inhibitory effects.

5. Cytochrome *c* is located on the outer leaflet of the inner membrane of the mitochondria closest to the intermembrane space. Damage to the outer membrane from any step of the PmFB prep can lead to partial loss of endogenous cytochrome *c*. Depending on the level of damage to the outer mitochondrial membrane, exogenous addition of cytochrome *c* will interact with the inner membrane and may stimulate respiration. In our experience, we have found that an increase in respiration by <15% after addition of cytochrome *c* is not associated with reductions in ADP-stimulated respiration, but this should be verified by each trained experimenter with numerous replicates in pilot experiments. Anecdotally, there is also an unresolved controversy as to whether the cytochrome *c* test may be indicative of outer membrane damage or permeability that existed in vivo in a given experimental model, as may be the case in certain disease states.

6. FCCP is a protonophore which increases permeability of the inner membrane to protons. This effect uncouples oxygen consumption at Complex IV from ATP synthesis at ATP synthase by drastically reducing the proton motive force [11, 12]. The increase in respiration detected after adding FCCP to the respirometry chamber does not reflect the degree to which mitochondria were uncoupled in their basal state. Rather, it is a test of the maximal rate at which electrons can be transported through the electron transport chain unimpeded by limitations of the phosphorylation system, which itself is defined as the transport of ADP/ATP across both mitochon drial membranes and ADP, Pi, and ATP cycling at ATP synthase. FCCP-stimulated rates of respiration that are greater than saturating ADP/Pi-stimulated respiration reflect a reserve capacity for electron flux in the ETC, whereas ADP/Pi-stimulated respiration reflects the functioning capacity of the phosphorylation system to support ATP synthesis. Caution is warranted to view FCCP-stimulated respiration as a measure of "maximal mitochondrial function" which is vague, non-specific, and can lead to misinterpretation of the actual ability of mitochondria to produce ATP.

7. Inhibitors which target specific sites on the ETC may help identify functional defects that may contribute to altered regulation of the ETC (Fig. 5). The loss of respiration in response to inhibitors can be interpreted in a deductive manner whereby the degree to which respiration declined is

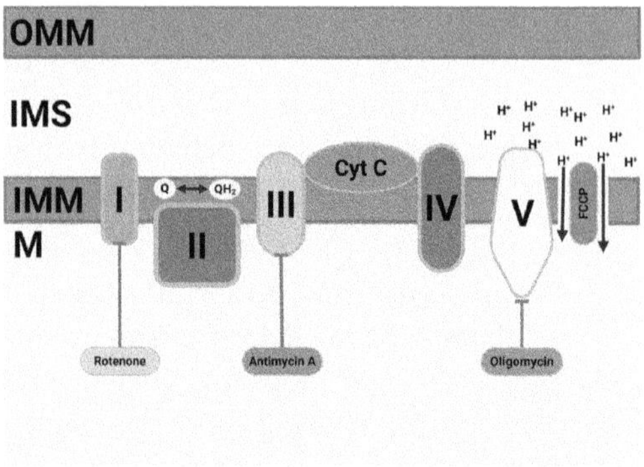

Fig. 5 Influence of rotenone, antimycin A, oligomycin, and FCCP on electron transport chain mechanics. Rotenone, antimycin A, and oligomycin inhibit Complex I (I), Complex III (III), and Complex V/ATP synthase (V), respectively. FCCP uncouples electron transport chain flux from ATP production at Complex V/ATP synthase by transporting protons from the intermembrane space (IMS) to the mitochondrial matrix (M). Complex II (II); complex IV (IV); inner mitochondrial membrane (IMM); outer mitochondrial membrane (OMM). Made using BioRender

reflective of the contribution of that component and downstream components of the ETC to the respiration rate that was monitored prior to the addition of the inhibitors. For example, if both NADH and $FADH_2$-generating substrates are used to stimulate respiration, subsequent additions of the Complex I inhibitor rotenone could identify the residual oxygen consumption that is due to $FADH_2$-stimulation of Complex II. Likewise, oligomycin-mediated inhibition of ATP synthase allows for assessment of respiration due to proton leak through the inner membrane, which is an indirect measure of uncoupled respiration. More direct measures of uncoupled respiration and proton leak should be considered if this is of interest in a given study.

8. Calibration ensures that the polarographic oxygen sensor measures the oxygen concentration and its rate of disappearance within the chamber as accurately as possible. This process can vary between brands and technologies. Therefore, it is important that the experimenter familiarizes themselves with the manufacturer's instructions and performs the calibration as frequently as required.

9. A major difference between MiR0 and Buffer Z includes the potassium (K^+) concentration. The [K^+] in MiR0 is 90 mM, whereas in Buffer Z, it is 145 mM, which is closer to the reported intracellular [K^+] of ~160 mM in human skeletal muscle [13] and 130–150 mM in rat skeletal muscle [14]. Other differences between buffers include the buffering agent (HEPES in MiR0 and K-MES in Buffer Z), presence of sucrose (only in MiRO) and taurine (only in MiR0)

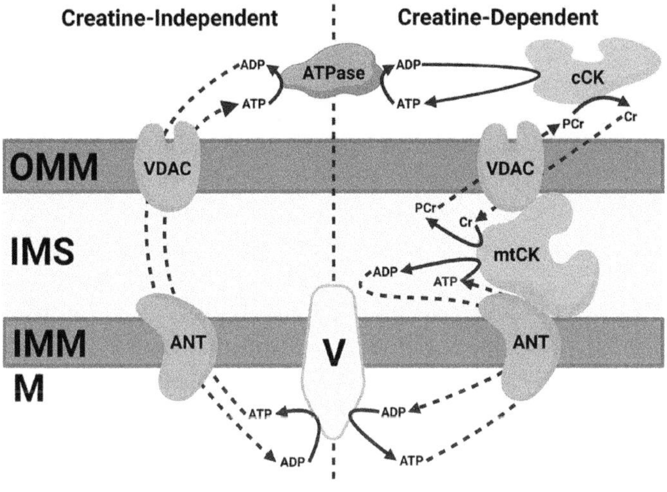

Fig. 6 Creatine-dependent and independent ATP transport pathways in skeletal muscle. Creatine and phosphocreatine diffuse across the mitochondrial membranes faster than ADP and ATP, respectively. For this reason, creatine accelerates and facilitates a "phosphate shuttling" mechanism from matrix to cytoplasmic ATP-utilizing proteins. Assessing respiration using both pathways can provide insight into these two predominant phosphate-transporting mechanisms as well as mitochondrial creatine kinase function. Inner mitochondrial membrane (IMM); intermembrane space (IMS); mitochondrial matrix (M); outer mitochondrial membrane (OMS). Made using BioRender

of which both are known to have antioxidant properties [15, 16] while the latter may also act as a buffering agent in the mitochondrial matrix [17]. However, a direct comparison between these two buffers in permeabilized HEK 293T cells and mouse muscle fiber bundles has not shown differences in mitochondrial respiration with certain protocols [18].

10. In order to simulate the in vivo environment as best as possible, temperature within the chamber is recommended to be kept at 37°C. Temperature may affect membrane stability, ATPase activity, and other intracellular processes, thus affecting rates of respiration and subsequent interpretation [19].

11. Transport of ATP across the two mitochondrial membranes occurs via creatine-dependent and independent mechanisms (Fig. 6) [20]. Presence of creatine (Cr) in the respiration buffer allows for creatine-dependent ATP shuttling, which utilizes mitochondrial (mtCK) and cytosolic (cCK) forms of creatine kinase. In this "phosphate shuttle," ATP is transported across the inner mitochondrial membrane (IMM) via ANT into the intermembrane space. Here, mtCK will transfer phosphate from ATP to creatine to produce phosphocreatine (PCr), which will then cross the outer mitochondrial membrane (OMM) via the voltage-dependent anion channel (VDAC) into the cytoplasm while ADP returns across

the inner membrane to the matrix. cCK will then transfer a phosphate from PCr to ADP to produce ATP, with creatine returning to the mitochondria. In this system, the theory predicts that ADP and ATP cycling occurs through short distances in both the mitochondria and cytoplasm near ATP-consuming proteins. This is predicted to accelerate the rate at which mitochondria produce ATP, and the rate at which ATP is synthesized locally at cytosolic ATP-consuming proteins, given ADP and ATP diffuse more slowly than creatine and phosphocreatine. In this way, phosphate is transferred across the longer diffusion distances between mitochondria and cytoplasmic ATP-consuming proteins by the faster creatine shuttle. Cr diffuses across the IMM and OMM approximately 2000 times faster than ADP, while PCr diffuses across approximately seven times faster than ATP, making creatine-independent shuttling of ATP faster than creatine -independent shuttling [21]. In contrast, the theory predicts that creatine-independent ATP shuttling involves direct diffusion of ATP via ANT and VDAC across the IMM and OMM. Therefore, mitochondrial respiration rates are often higher when creatine is included during respirometry, particularly at lower ADP concentrations (≤500 mM), given creatine has a unique effect of increasing mitochondrial ADP sensitivity. Simultaneous respiration experiments in the presence and absence of creatine kinase in separate permeabilized fiber bundles can be used to screen for potential mtCK defects or enhancements in experimental models [22]. To saturate and maximally stimulate mtCK, add creatine at a concentration of 20 mM [23–25]. This system can also be assessed by modeling both creatine and phosphocreatine concentrations, which have opposing influences on mtCK [25, 26]. Both creatine and phosphocreatine can be added the day of the experiment in a separate aliquot of the respiration medium, as stability can be uncertain in cold storage. Mixing should be performed gently on a nutator or rotator for ~5 min to avoid bubbles.

12. Several commonly used anesthetics have been found to impair mitochondrial function [27]. Specifically, isoflurane has been found to generate mitochondrial ROS after 6 h of exposure [28]. The goal should be to reduce time to anesthetic exposure as much as possible; therefore, we recommend that tissue harvest be completed within approximately 30 min.
13. Lidocaine can inhibit mitochondrial function and induce mitochondrial cell death pathways [29, 30]. Therefore, it is very important to avoid injecting lidocaine into muscle.
14. Micropunch biopsy technique involves the use of a 12-gauge cannula which guides a needle into muscle to a depth of

around 4 cm and makes about four to five cuts [31]. The Bergstrom technique requires a small, superficial incision on the skin to allow a hollow needle with a small opening to pierce the fascia and skeletal muscle underneath [32]. A cylinder with sharp edges is then inserted into the hollow needle to make about three to four cuts of the muscle within the window of the needle, and a hollow syringe is used to suck the pieces up the needle. The Bergstrom technique allows you to obtain longer and more connected fibers, while microbiopsy fibers tend to be shorter and less connected [31]. A previous investigation found that in the absence of blebbistatin in the respiration buffer, microbiopsy PmFBs tended to disintegrate more and showed lower glutamate-supported respiration compared with Bergstrom PmFBs [31]. However, the presence of blebbistatin in the buffer abolished these differences, but succinate-supported respiration was higher in the Bergstrom fibers. In addition, it was also reported that muscle obtained using microbiopsy technique was composed of more type IIx fibers and fewer type I fibers compared to Bergstrom. Although both techniques are valid, choosing the best one will depend on the study design and experimental question.

15. To obtain the best possible visibility, place a dark background underneath the dish and have a light source pointed at the tissue. The frozen ice-pack should also be changed if separating for longer than 10 min to ensure that the muscle is kept cold.
16. **Longitudinal stress warning:** avoid using a sliding motion when cutting, as that may cause longitudinal stress on the fibers leading to damage of the mitochondrial reticulum.
17. **Longitudinal stress warning:** it is important to perform this step using very small, controlled movements so as to avoid damaging the mitochondrial reticulum.
18. Learning to identify optimal separation is an observational skill that comes with plenty of practice. The degree of separation can heavily influence the oxygen consumption rates; therefore, during practice, it is important to run respirometry on several bundles and compare the rates to literature as an indicator of one's skill.
19. Generally speaking, if a bundle is too small, the level of total oxygen consumption across the mitochondrial network may not be high enough for the polarographic oxygen sensor to detect. Conversely, if a bundle is too big, the mitochondrial network may consume oxygen very rapidly. In this scenario, the chamber oxygen concentration may decline to a level that becomes rate-limiting for respiration. As noted above, reoxygenating the chamber can restore adequate oxygen

concentrations but may result in inconsistent measurements of respiration rates. Optimal bundle sizes also vary between skeletal muscle types due to the differences in mitochondrial content. For skeletal muscles with less mitochondrial content [33], optimal bundle sizes will be larger compared to the size required for more oxidative muscles with higher mitochondrial content [33]. Furthermore, the type of substrates used within a protocol (e.g., pyruvate-malate, palmitoyl-CoA) and the animal species used to obtain a given muscle type are also determinants of optimal bundle size. The experimenter should run a few pilot bundles of various sizes for a given muscle and protocol to determine their own optimal bundle sizes.

20. **Longitudinal stress warning:** when removing bundles from tissue wipes, be sure to use a scooping motion with the forceps to pick up the whole bundle at once. Avoid leaving the bundle on the tissue wipe for more than a few seconds.

21. Saponin permeabilizes membranes by selectively interacting with cholesterol [34]. Plasma membranes have high cholesterol content; therefore, saponin is very effective in permeabilizing this layer [35]. Mitochondrial membranes have much lower cholesterol content and are only affected at very high saponin concentrations [35]. Unpublished observations have led us to determine that saponin concentrations of 40 μg/mL and 30 μg/mL are best for rodent and human skeletal muscle, respectively. These concentrations yield the highest rates without compromising the integrity of the outer mitochondrial membrane.

22. Always ensure that there is an air bubble in the tube that rocks back and forth with the nutating motion. This method ensures that the bundle is constantly moving across the length of the tube and all fibers are uniformly exposed to the buffer (BIOPS and saponin/Buffer Z/MiRO).

23. The high degree of sensitivity afforded by high resolution respirometry allows the experimenter an opportunity to design protocols that simulate potential scenarios that the mitochondria may encounter *in vivo*. Generally, respirometry protocols involve the titration of compounds in the following order (*see* **Note 26**):

    i) Substrates which generate reducing equivalents in the form of nicotinamide adenine dinucleotide (NADH) and flavin adenine dinucleotide ($FADH_2$) (*see* **Note 27**).

    ii) Adenosine diphosphate (ADP) to allow for adenosine triphosphate (ATP) generation (*see* **Note 28**).

    iii) Inhibitors which target specific sites on the electron transport chain (*see* **Note 7**).

iv) Cytochrome *c* as a quality control test of outer-mitochondrial membrane integrity (*see* **Note 5**).

24. High resolution respirometry permits the measurement of oxygen consumption as a rate, or oxygen flux, calculated as a derivation of the change (slope) of oxygen concentration over time (Fig. 4). While all respirometers measure changes in oxygen concentration within the chamber, some manufacturers provide real-time calculations of the flux derivation while others may require such calculations after the experiment is completed.

25. Respiration rates can be normalized to bundle wet weight or dry weight. This measure will be influenced by the quantity of mitochondria per mg weight as well as the intrinsic properties of the mitochondria. Respiration rates normalized per mg weight can be normalized a second time to markers of mitochondrial content to provide insight into intrinsic properties of mitochondria independent of mitochondrial content. A comparison of both approaches can provide important insight into whether mitochondrial content or intrinsic properties exist between experimental groups in a given study. For example, if pyruvate oxidation was different between two groups when normalized to mg weight as well as per marker of mitochondrial content, then it is possible that the intrinsic regulation of pyruvate metabolism has been altered. An example follow-up experiment could include measurement of pyruvate dehydrogenase phosphorylation state as a marker of post-translational regulation of this pathway. Also, when normalizing to mitochondrial content markers, the experimenter could also consider normalizing respiration rates to the content of specific proteins that were stimulated in a given protocol. For example, normalizing respiration rates to the sum of several electron transport chain protein contents assessed by western blot may be relevant given ADP-stimulated respiration is a result of flux through this system regardless of the substrate used to generate reducing equivalents. These examples might inspire the experimenter to reflect on how the design of substrate titration protocols guides the interpretation of results to specific proteins that may be altered in a given study.

26. This order of substrate titration can allow for the determination of ADP sensitivity and the maximum respiration rate for the specific substrate used to produce reducing equivalents. However, this order can be switched to determine the sensitivity of the mitochondria to a specific reducing equivalent-generating substrate (e.g., pyruvate). First, a saturating bolus of ADP along with other supporting substrates (e.g.,

malate) is added, followed by gradual titrations of the substrate of interest. Sensitivity measures should ideally be performed with consideration of principles used to calculate the apparent Km of the titrated substrate with assurances that the modeling approach provides a strong goodness of fit (e.g., Michaelis–Menten analyses, other approaches) [36].

27. In the absence of ADP, this will provide a leak respiration rate, which is an indicator of uncoupled mitochondrial respiration.
28. This approach will assess oxidative phosphorylation capacity of the reducing equivalent generating substrate in step 1 by adding a saturating bolus of ADP (>5 mM). Lower concentrations (25–500 μM) can be added to assess ADP sensitivity as well.

References

1. Gnaiger E (2008) Polarographic oxygen sensors, the oxygraph and high-resolution respirometry to assess mitochondrial function. In: Drug-induced mitochondrial dysfunction, pp 327–352
2. Schmidt CA, Fisher-Wellman KH, Neufer PD (2021) From OCR and ECAR to energy: perspectives on the design and interpretation of bioenergetics studies. J Biol Chem 297(4):101140
3. Gnaiger E (2007) Isolated mitochondria or permeabilized tissues and cells. 11(983):1–5
4. Pesta D, Gnaiger E (2012) High-resolution respirometry: OXPHOS protocols for human cells and permeabilized fibers from small biopsies of human muscle. Methods Mol Biol 810:25–58
5. Picard M, Taivassalo T, Ritchie D, Wright KJ, Thomas MM, Romestaing C et al (2011) Mitochondrial structure and function are disrupted by standard isolation methods. PLoS One 6(3):e18317
6. Pan JW, Hamm JR, Rothman DL, Shulman RG (1988) Intracellular pH in human skeletal muscle by 1H NMR. Proc Natl Acad Sci U S A 85(21):7836–7839
7. Chen YW, Liu HQ, Wu Q, Huang YH, Tung YY, Lin MH et al (2020) pH mapping of skeletal muscle by chemical exchange saturation transfer (CEST) imaging. Cells 9(12):2610
8. Allsop P, Cheetham M, Brooks S, Hall GM, Williams C (1990) Continuous intramuscular pH measurement during the recovery from brief, maximal exercise in man. Eur J Appl Physiol Occup Physiol 59(6):465–470
9. Street D, Bangsbo J, Juel C (2001) Interstitial pH in human skeletal muscle during and after dynamic graded exercise. J Physiol 537(Pt 3):993–998
10. Perry CG, Kane DA, Lin CT, Kozy R, Cathey BL, Lark DS et al (2011) Inhibiting myosin-ATPase reveals a dynamic range of mitochondrial respiratory control in skeletal muscle. Biochem J 437(2):215–222
11. Benz R, McLaughlin S (1983) The molecular mechanism of action of the proton ionophore FCCP (carbonylcyanide p-trifluoromethoxyphenylhydrazone). Biophys J 41(3):381–398
12. Zorova LD, Popkov VA, Plotnikov EY, Silachev DN, Pevzner IB, Jankauskas SS et al (2018) Mitochondrial membrane potential. Anal Biochem 552:50–59
13. Sjogaard G, Adams RP, Saltin B (1985) Water and ion shifts in skeletal muscle of humans with intense dynamic knee extension. Am J Physiol 248(2 Pt 2):R190–R196
14. Nagaoka R, Yamashita S, Mizuno M, Akaike N (1994) Intracellular Na+ and K+ shifts induced by contractile activities of rat skeletal muscles. Comp Biochem Physiol A Physiol 109(4):957–965
15. Jong CJ, Sandal P, Schaffer SW (2021) The role of taurine in mitochondria health: more than just an antioxidant. Molecules 26(16):4913
16. Xu YJ, Arneja AS, Tappia PS, Dhalla NS (2008) The potential health benefits of taurine in cardiovascular disease. Exp Clin Cardiol 13(2):57–65
17. Hansen SH, Andersen ML, Cornett C, Gradinaru R, Grunnet N (2010) A role for taurine in mitochondrial function. J Biomed Sci 17(Suppl 1):S23
18. Komlodi T, Sobotka O, Krumschnabel G, Bezuidenhout N, Hiller E, Doerrier C et al (2018) Comparison of mitochondrial incubation media for measurement of respiration and hydrogen peroxide production. Methods Mol Biol 1782:137–55.

19. Perry CG, Kane DA, Lanza IR, Neufer PD (2013) Methods for assessing mitochondrial function in diabetes. Diabetes 62(4):1041–1053
20. Wallimann T, Tokarska-Schlattner M, Schlattner U (2011) The creatine kinase system and pleiotropic effects of creatine. Amino Acids 40(5):1271–1296
21. Kaldis P, Kamp G, Piendl T, Wallimann T (1997) Functions of creatine kinase isoenzymes in spermatozoa. Adv Dev Biol (1992) 5:275-312
22. Hughes MC, Ramos SV, Brahmbhatt AN, Turnbull PC, Polidovitch NN, Garibotti MC et al (2024) Mitohormesis during advanced stages of Duchenne muscular dystrophy reveals a redox-sensitive creatine pathway that can be enhanced by the mitochondrial targeting peptide SBT-20. Redox Biol 76:103319
23. Perry CG, Kane DA, Herbst EA, Mukai K, Lark DS, Wright DC et al (2012) Mitochondrial creatine kinase activity and phosphate shuttling are acutely regulated by exercise in human skeletal muscle. J Physiol 590(21):5475–5486
24. Ydfors M, Hughes MC, Laham R, Schlattner U, Norrbom J, Perry CG (2016) Modelling in vivo creatine/phosphocreatine in vitro reveals divergent adaptations in human muscle mitochondrial respiratory control by ADP after acute and chronic exercise. J Physiol 594(11):3127–3140
25. Hughes MC, Ramos SV, Turnbull PC, Rebalka IA, Cao A, Monaco CMF et al (2019) Early myopathy in Duchenne muscular dystrophy is associated with elevated mitochondrial H(2) O(2) emission during impaired oxidative phosphorylation. J Cachexia Sarcopenia Muscle 10(3):643–661
26. Perry CG, Heigenhauser GJ, Bonen A, Spriet LL (2008) High-intensity aerobic interval training increases fat and carbohydrate metabolic capacities in human skeletal muscle. Appl Physiol Nutr Metab 33(6):1112–1123
27. La Monaca E, Fodale V (2012) Effects of anesthetics on mitochondrial signaling and function. Curr Drug Saf 7(2):126–139
28. Zhang Y, Dong Y, Wu X, Lu Y, Xu Z, Knapp A et al (2010) The mitochondrial pathway of anesthetic isoflurane-induced apoptosis. J Biol Chem 285(6):4025–37.
29. Tan X, Liu R, Dan L, Huang H, Duan C (2024) Effects of anesthetics on mitochondrial quality control: mechanisms and clinical implications. Anesthesiology and Perioperative. Science. 2(3):31
30. Kawasaki C, Kawasaki T, Ogata M, Sata T, Chaudry IH (2010) Lidocaine enhances apoptosis and suppresses mitochondrial functions of human neutrophil in vitro. J Trauma 68(2):401–408
31. Hughes MC, Ramos SV, Turnbull PC, Nejatbakhsh A, Baechler BL, Tahmasebi H et al (2015) Mitochondrial bioenergetics and fiber type assessments in microbiopsy vs. bergstrom percutaneous sampling of human skeletal muscle. Front Physiol 6:360
32. Bergstrom J (1975) Percutaneous needle biopsy of skeletal muscle in physiological and clinical research. Scand J Clin Lab Invest 35(7):609–616
33. Dong H, Tsai SY (2023) Mitochondrial properties in skeletal muscle fiber. Cells 12(17):2183
34. Bangham AD, Horne RW, Glauert AM, Dingle JT, Lucy JA (1962) Action of saponin on biological cell membranes. Nature 196:952–955
35. Goicoechea L, Conde de la Rosa L, Torres S, Garcia-Ruiz C, Fernandez-Checa JC (2023) Mitochondrial cholesterol: metabolism and impact on redox biology and disease. Redox Biol 61:102643
36. Zuccarelli L, De Martino M, Filippi A, Knapton AE, Thackray BD, Baldassarre G et al (2025) Mitochondrial sensitivity to submaximal [ADP] following bed rest: a novel two-phase approach associated with fibre types. J Cachexia Sarcopenia Muscle 16(3):e13775
37. Brand MD, Nicholls DG (2011) Assessing mitochondrial dysfunction in cells. Biochem J 435(2):297–312
38. Kuznetsov AV, Veksler V, Gellerich FN, Saks V, Margreiter R, Kunz WS (2008) Analysis of mitochondrial function in situ in permeabilized muscle fibers, tissues and cells. Nat Protoc 3(6):965–976
39. Ogata T, Yamasaki Y (1997) Ultra-high-resolution scanning electron microscopy of mitochondria and sarcoplasmic reticulum arrangement in human red, white, and intermediate muscle fibers. Anat Rec 248(2):214–223
40. Doerrier C, Garcia-Souza LF, Krumschnabel G, Wohlfarter Y, Meszaros AT, Gnaiger E (2018) High-resolution FluoRespirometry and OXPHOS protocols for human cells, permeabilized fibers from small biopsies of muscle, and isolated mitochondria. Methods Mol Biol 1782:31–70
41. Larsen S, Kraunsoe R, Gram M, Gnaiger E, Helge JW, Dela F (2014) The best approach: homogenization or manual permeabilization of human skeletal muscle fibers for respirometry?. Anal Biochem 446:64–68
42. Gnaiger E (2003) Oxygen conformance of cellular respiration: a perspective of mitochondrial physiology. In: Hypoxia: through the lifecycle, pp 39–55.
43. Eigentler A, Fontana-Ayoub M, Gnaiger E (2013) O2k-Fluorometry: HRR and H2O2 production in mouse cardiac tissue homogenate. Mitochondr Physiol Network 18:1–6
44. Djafarzadeh S, Jakob SM (2017) High-resolution respirometry to assess mitochondrial function in permeabilized and intact cells. J Vis Exp 120:54985

Check for updates

Chapter 17

Applying Surface Plasmon Resonance to the Study of Striated Muscle Protein Interactions

Sabrina Apel, SaiLavanyaa Sundar, Jeffrey Moore and Matthew Gage

Abstract

Surface Plasmon Resonance (SPR) is a powerful tool for measuring binding interactions, especially in biological systems where many interactions have binding affinities in the nanomolar to micromolar range. SPR has been used to investigate binding interactions in a wide range of systems but has largely not been used to explore striated muscle protein interactions. This chapter outlines the method that our labs have used to measure binding interactions involving striated muscle proteins. The first method focuses on binding of a calcium-binding protein to the titin protein in a traditional SPR binding experiment. The second method focuses on the multi-site binding of truncated fragments of the regulatory protein troponin T (TnT1) to the tropomyosin-filamentous actin (Tpm/F-actin) thin filament complex, which involves multiple interaction sites. These two approaches highlight the utility of using SPR to understand muscle protein interactions.

Keywords Surface plasmon resonance, S100A1, titin, UN2A, F-actin, Troponin, Tropomyosin

1 Introduction

Surface Plasmon Resonance (SPR) is a powerful, label-free method to measure binding interactions, especially between two proteins, in real-time [1–4]. One protein (commonly referred to as the ligand) is attached to the sensor using a variety of chemical strategies, often optimized to ensure active and proper orientation of the ligand [5, 6]. The second protein, the analyte, is flowed across the surface at varying concentrations [1]. When binding occurs, it leads to a change in the refractive index at the sensor–analyte interface, producing a resonance angle shift that yields a real-time SPR response [7, 8]. The signal change depends on both the amount of mass attached and the molecular size of the analyte [8]. An overview of the basic SPR workflow is shown in Fig. 1. One of the key strengths of SPR is its ability to measure both association and dissociation kinetics, as well as steady-state binding, providing a comprehensive view of interaction dynamics [1, 2].

Joe Quadrilatero (ed.), *Myogenesis and Skeletal Muscle Biology*, Methods in Molecular Biology, vol. 3057,
https://doi.org/10.1007/978-1-0716-5396-8_17,

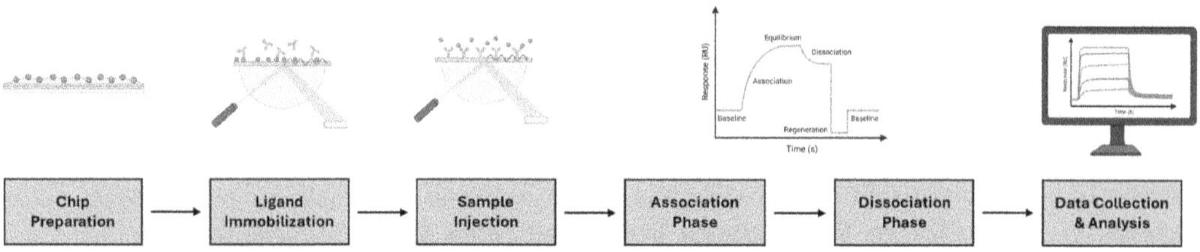

Fig. 1 Workflow of surface plasmon resonance (SPR) experiments. This schematic depicts the main stages of an SPR experiment. After chip preparation and ligand immobilization, analyte samples are flowed across the sensor surface. Binding is monitored in real time during the association and dissociation phases, and the sensorgrams are subsequently analyzed to determine kinetic and equilibrium parameters. Figure prepared using Biorender

SPR works based on total internal reflection (TIR) of a light beam in a prism—typically a glass prism—under conditions where the light propagates in a denser medium and meets a boundary with a less dense medium above a critical angle [9]. A plane-polarized light beam is directed through the prism and reflects off a thin metallic film, usually a ~50 nm layer of gold, deposited on the prism surface [9, 10]. When the light reaches the correct resonance angle, the energy is transferred to the electrons in the metal, exciting surface plasmon-polaritons (SPPs) and producing an evanescent wave at the metal–dielectric interface. As molecules bind to the metal surface, they alter the local refractive index in the evanescent field region, which causes a measurable shift in the resonance angle [11].

Proteins of interest can be coupled to the gold surface using a variety of different types of chemistries. The most common attachment is either through carboxyl- or amine-mediated covalent coupling, typically using EDC/NHS activation of carboxymethyl dextran sensor surfaces [12, 13]. However, chips enabling thiol-based (disulfide) coupling, biotin-streptavidin, and even His-tag/Ni^{2+}chelation have been developed [12]. These affinity-based strategies allow for controlled and oriented attachment, unlike the more random orientation often observed with general carboxyl and amine coupling.

There are three kinetic phases to bind during an SPR experiment. The initial stage, when the analyte is initially injected, is the association phase, where the response increases due to binding to the ligand [14, 15]. Once the binding progresses sufficiently, it enters the steady-state phase, where rates of association and dissociation are balanced, resulting in a plateau in the response signal [16]. The final stage, the dissociation phase, begins once the analyte flow is stopped and only the buffer flows, causing the bound analyte to dissociate, leading to a decline in response [15]. These kinetic parameters: the association rate constant (k_a), dissociation rate constant (k_d), and the equilibrium dissociation constant (K_D), can be used to characterize the binding thermodynamics according to various binding models [14, 16].

The first BIAcore instrument, one of the workhorses in the SPR field, was commercially released in 1990 by Pharmacia Biosensor AB and marked a pivotal moment in transforming SPR into a robust analytical tool for biomolecular interaction studies [17, 18]. Since that time, the number of companies manufacturing SPR instruments has expanded considerably, with estimates suggesting around 25 suppliers have emerged over the years. One of the most affordable instruments available today is the OpenSPR from Nicoya, which employs localized SPR (LSPR) via nanoparticles rather than a continuous gold film [19]. This was the instrument used for the two methods described below.

2 Materials

2.1 Materials for Binding of S100A1 to the UN2A Region of Titin

1. Purified recombinantly expressed S100A1 protein.
2. Purified recombinantly expressed UN2A protein.
3. 2-channel OpenSPR instrument (Nicoya Lifesciences) with 100 μM sample loop.
4. OpenSPR™ High Capacity Carboxyl Sensor chips.
5. OpenSPR™ Carboxyl Reagent Kit, including EDC/NHS coupling solution and Ethanolamine blocking solution.
6. Calcium-free solution: 250 mM NaCl, 20 mM HEPES, 0.05% Tween, 1 mM EGTA.
7. Calcium-binding solution: 250 mM NaCl, 20 mM HEPES, 0.05% Tween, and varying $CaCl_2$ concentrations to reach pCa 7, 5, and 3. Solution pH was adjusted to either pH 6, 6.5, or 7.4.
8. Regeneration buffer: 250 mM NaCl, 20 mM HEPES, 1 mM EDTA.

2.2 Materials for Binding of TnT1 to Tropomyosin-Actin Complexes

1. Purified biotin-conjugated G-actin.
2. Purified recombinantly expressed tropomyosin.
3. Purified recombinantly expressed TnT1 fragment 1-156.
4. Purified recombinantly expressed TnT1 fragment 89-156.
5. Purified recombinantly expressed TnT1 fragment 100-156.
6. 2-channel OpenSPR instrument (Nicoya Lifesciences) with 100 μM sample loop.
7. OpenSPR™ Biotin-Streptavidin Sensor chips.
8. Biotinylated skeletal muscle actin (Cytoskeleton, Cat # AB07-A).
9. Running Buffer: 1× PBS (pH 7.4), 1 mg/mL BSA, 0.05% Tween-20.
10. G-Actin buffer: 5 mM Tris–HCl (pH 8.0), 0.2 mM $CaCl_2$, 0.2 mM ATP, 0.5 mM DTT.

11. F-Actin buffer: 1 M Tris–HCl (pH 8.0), 2.0 M KCl, 1.0 M $MgCl_2$, 0.1 M ATP.

3 Methods

Methods for Binding of S100A1 to the UN2A Region of Titan

3.1 Sensor Conditioning and Ligand Immobilization

1. Remove new chip from the storage container and gently wash with ddH_2O.
2. Dry the surface of the chip carefully with compressed air (*see* **Note 1**).
3. Place the sensor chip into the OpenSPR and press the load button on the software to seat the chip into the instrument.
4. Load 10 mM HCl solution into the sample loop and then flow across the sensor chip at a flow rate of 150 μL/min using the conditioning step on the software. There should be a drop in the signal as the HCl cleans the surface, and then the signal will return to a steady state.
5. Once the signal returns to a steady state, load freshly prepared EDC/NHS solution into the sample loop and apply the EDC/NHS to the chip at a 20 μL/min flow rate. This step will cause binding of the crosslinker to the chip in preparation for binding the ligand
6. Flow the ligand (purified UN2A protein diluted in sodium acetate to 10–50 μg/mL) over the activated chip at a flow rate of 20 μL/min. The target signal should be 66 RU, which is calculated based on the anticipated signal based on the size of the ligand (*see* **Note 2**).
7. The surface is then blocked using Ethanolamine, which binds to any unreacted active sites to prevent binding of the analyte to the chip. The ethanolamine blocking solution is loaded into the sample loop and flowed across the chip at a flow rate of 20 μL/ml.

3.2 Preparation of Analyte Sample

1. Prepare 3–4 liters of binding buffer for the given set of conditions.
2. Place 1 mL of S100A1 into a dialysis membrane and place the loaded membrane into 1 L of binding buffer in a beaker
3. Incubate the solution at 4°C for 4–24 h to allow the solution to reach equilibrium.
4. Exchange the solution and allow the solution to reach equilibrium again. Repeat the process one to two more times.
5. Remove the protein from the dialysis membrane after the final equilibration step.

6. Determine protein concentration by either Bradford or Nanodrop and dilute protein to the desired concentration (5–100 μM) for the binding assay.
7. Collect the final equilibration dialysate to use as running buffer for the binding assay (*see* **Note 3**).

3.3 Binding Assay

1. Load sample into sample loop and run the sample injection program. This will inject the sample over a 2-min period (association phase) followed by a 5-min dissociation phase.
2. The surface is regenerated using 20 mM HEPES, 250 mM NaCl, and 1 mM EDTA. The EDTA removes the calcium bound to the S100A1, causing it to dissociate from the UN2A ligand.
3. Repeat these steps for the various protein concentrations of interest.
4. For additional solution conditions, Step 3.2 needs to be repeated for each buffer condition.

3.4 Data Analysis

1. Sensorgrams are imported into the TraceDrawer™ Analysis software package that is provided by Nicoya for their OpenSPR.
2. Using a two-channel chip, the baseline is automatically subtracted from the sensorgram. However, in the case of a single channel instrument, background sensorgrams are collected for each condition to correct for non-specific binding. These are

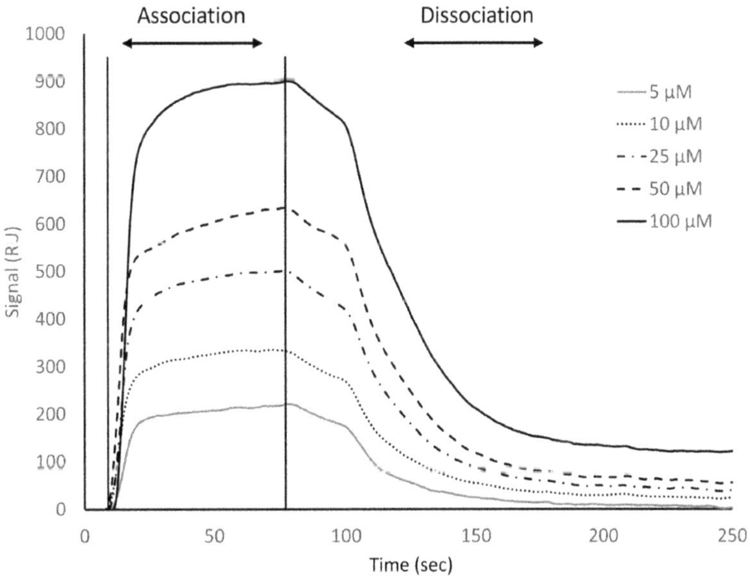

Fig. 2 Representative data from S100A1/UN2A binding. S100A1 concentration is shown on the right. The association stage is highlighted to show where the S100A1 was being flowed across the surface. (©2024 Apel et al. Originally published in Journal of General Physiology. https://doi.org/10.1085/jgp.202313472)

subtracted from the experimental sensorgram to correct for non-specific binding.

3. Traces are fit to a 1:1 binding model to determine the on and off binding rates (Fig. 2).

Methods for Binding of TnT1 to Tropomyosin-Actin Complexes

3.5 Preparation of F-Actin Solution

1. Resuspend lyophilized biotin-actin in 20 μL G-actin buffer (final concentration 23 μM) and incubate on ice for 30 min.
2. Centrifuge at 18,000 × *g* for 15 min at ~22°C to remove aggregated G-actin, retain the supernatant.
3. Polymerize the G-actin to form F-actin by diluting G-actin ten-fold (20 μL into 180 μL) in F-actin buffer.
4. Centrifuge at 140,000 × *g* for 30 min at 4°C.
5. Resuspend the pelleted F-actin into 400 μL of Running Buffer (final concentration 1.3 μM).

3.6 Sensor Conditioning and Ligand Immobilization

1. Remove new chip from the storage container and gently wash with ddH_2O.
2. Dry the surface of the chip carefully with compressed air.
3. Place the sensor chip into the OpenSPR and press the load button on the software to seat the chip into the instrument.
4. Load 10 mM HCl solution into the sample loop and then flow across the sensor chip at a flow rate of 150 μL/min using the conditioning step on the software. There should be a steep decline in the signal as the HCl cleans the surface, and then the signal will return to a steady state (Fig. 3).
5. Once the signal returns to a steady state, prepare a 0.5 μM streptavidin solution in running buffer.

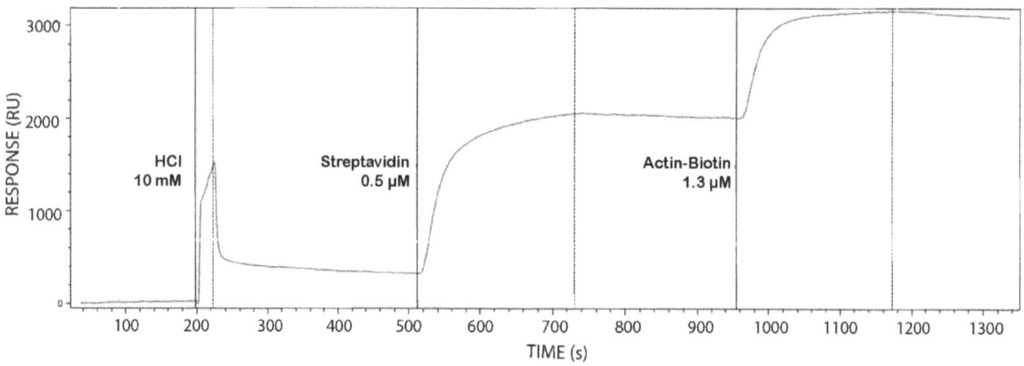

Fig. 3 Representative ligand immobilization sensorgram showing time vs. response (in response units, RU). Sensorgram depicting the SPR response for the sensor conditioning (using 10 mM HCl), followed by sequential immobilization of 0.5 μM streptavidin and 1.3 μM biotin-actin, before addition of 4.0 μM Tpm, to form the stable Tpm-actin ligand complex

6. Flow the 0.5 μM streptavidin solution over the activated chip at a flow rate of 20 μL/min. Incubate for 5 min to allow the biotin to coat the surface of the sensor.
7. Load previously prepared biotin–actin solution into the sample loop and flow across the surface of the sensor for 5 min so that the F-actin is captured by streptavidin.
8. Wash weakly bound F-actin by injecting running buffer over the sensor at a flow rate of 200 μL/min.

3.7 Binding of Tpm to Stable Ligand Complex (F-Actin/Tpm Complex)

1. Dilute Tpm sample to 4.0 μM in running buffer.
2. Incubate at 22°C for 10 min prior to injection to avoid thermal effects.
3. Load sample into sample loop and inject at 20 μL/min.
4. Wash with running buffer at 200 μL/min to remove unbound Tpm and to stabilize the complex.
5. Perform additional injections of the Tpm solution until the signal reaches a steady state, including saturation of Tpm-actin binding.

3.8 Preparation of Analyte Sample

1. Dialyze TnT1 fragments (1–156, 89–156, or 100–156) into Running Buffer.
2. Load sample into sample loop and inject at 20 μL/min to allow binding of Tpm and F-actin immobilized on the sensor.
3. Samples are injected from low to high concentrations (1–15 μM) to avoid any residual bound TnT1 from impacting the measured binding curves.
4. No sensor regeneration is required for this experiment since the analyte would completely dissociate (*see* **Note 4**).

3.9 Data Analysis

1. Sensorgrams are imported into the TraceDrawer™ Analysis software package that is provided by Nicoya for their OpenSPR.
2. Using a two-channel chip, the baseline is automatically subtracted from the sensorgram. However, in the case of a single channel instrument, background sensorgrams are collected for each condition to correct for non-specific binding. These are subtracted from the experimental sensorgram to correct for non-specific binding (*see* **Note 5**).
3. Traces are fit to a two-state kinetic model for TnT1-156 binding to the Tpm/F-actin ligand complex since the sensorgrams depict fast and slow association and dissociation phases with two interaction sites (Fig. 4A).

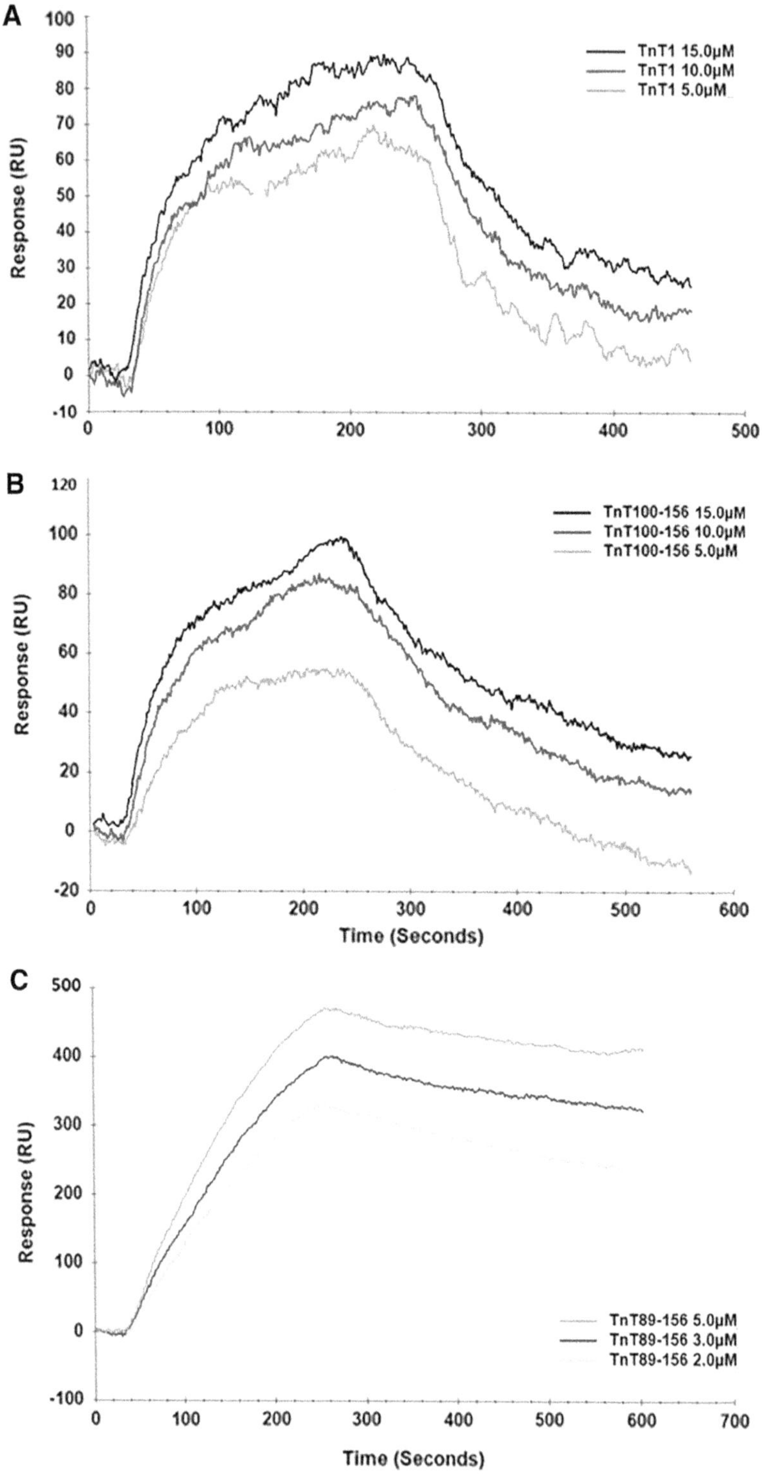

Fig. 4 Representative real-time association and dissociation response of multiple concentrations of TnT1-156, TnT100-156, and TnT89-156 fragments interacting with immobilized Tpm-actin. (**A**) TnT1-156 fragment exhibits a two-site interaction model with a fast and slow association and dissociation phases; (**B**) one-site

4. For binding of truncated fragments, TnT100-156 and TnT89-156 to the Tpm/F-actin complex, traces are fit to a 1:1 binding model since the data clearly represent a one-site interaction (Figs. 4B and C).

4 Notes

1. Be sure to completely dry the sensor chip or it won't seat properly, and you will get leaks in the microfluidics.
2. The reference channel should not receive any ligand so that it can provide a measure of non-specific binding to the chip.
3. It is extremely important that the analyte is dialyzed into the same buffer used as the mobile phase to avoid changes in the refractive index due to buffer-buffer interactions.
4. No sensor regeneration is required for the TnT1-Tpm/F-actin binding experiment since the TnT1 dissociated completely during the experiment.
5. TnT1 non-specific binding to F-actin is determined by measuring the sensorgram when the TnT1 analyte was flowed across attached F-actin without Tpm bound.

References

1. Douzi B (2017) Protein–protein interactions: surface plasmon resonance. In: Journet L, Cascales E (Eds.) Bacterial protein secretion systems: methods and protocols. Springer, New York, NY, 257–275
2. Homola J (2008) Surface plasmon resonance sensors for detection of chemical and biological species. Chem Rev 108(2):462–493
3. Myszka DG, He X, Dembo M, Morton TA, Goldstein B (1998) Extending the range of rate constants available from BIACORE: interpreting mass transport-influenced binding data. Biophys J 75(2):583–594
4. Nikolovska-Coleska Z (2015) Studying protein-protein interactions using surface plasmon resonance. In: Meyerkord CL, Fu H (Eds.) Protein-protein interactions: methods and applications. Springer, New York, NY, 109–138
5. Lahiri J, Isaacs L, Tien J, Whitesides GM (1999) A strategy for the generation of surfaces presenting ligands for studies of binding based on an active ester as a common reactive intermediate: a surface plasmon resonance study. Anal Chem 71(4):777–790
6. Löfås S, Johnsson B (1990) A novel hydrogel matrix on gold surfaces in surface plasmon resonance sensors for fast and efficient covalent immobilization of ligands. J Chem Soc Chem Commun 21:1526–1528
7. Rich RL, Myszka DG (2000) Advances in surface plasmon resonance biosensor analysis. Curr Opin Biotechnol 11(1):54–61
8. Stenberg E, Persson B, Roos H, Urbaniczky C (1991) Quantitative determination of surface concentration of protein with surface plasmon resonance using radiolabeled proteins. J Colloid Interface Sci 143(2):513–526

Fig. 4 (continued) binding of the TnT100-156 fragment to Tpm-actin; (**C**) TnT89-156 exhibits one-site binding with low dissociation from the Tpm-actin complex. Experiments were performed in triplicates after stable immobilization of Tpm-actin on the streptavidin-modified surface

9. Markey F (2000) Principles of surface plasmon resonance. In: Nagata K, Handa H (Eds.) Real-time analysis of biomolecular interactions: applications of BIACORE. Springer Japan, Tokyo, 13–22
10. Akimov Y (2018) Optical resonances in Kretschmann and Otto configurations. Opt Lett 43(6):1195–1198
11. Piliarik M, Homola J (2009) Surface plasmon resonance (SPR) sensors: approaching their limits?. Opt Express 17(19):16505–16517
12. Hutsell SQ, Kimple RJ, Siderovski DP, Willard FS, Kimple AJ (2010) High-affinity immobilization of proteins using biotin- and GST-based coupling strategies. Methods Mol Biol 627:75–90
13. Johnsson B, Lofas S, Lindquist G (1991) Immobilization of proteins to a carboxymethyl-dextran-modified gold surface for biospecific interaction analysis in surface plasmon resonance sensors. Anal Biochem 198(2):268–277
14. Karlsson R, Michaelsson A, Mattsson L (1991) Kinetic analysis of monoclonal antibody-antigen interactions with a new biosensor based analytical system. J Immunol Methods 145(1-2):229–240
15. Oshannessy DJ, Brighamburke M, Soneson KK, Hensley P, Brooks I (1993) Determination of rate and equilibrium binding constants for macromolecular interactions using surface plasmon resonance: use of nonlinear least squares analysis methods. Anal Biochem 212(2):457–468
16. Navratilova I, Myszka DG (2006) Investigating biomolecular interactions and binding properties using SPR biosensors. In: Homola J (Ed.) Surface plasmon resonance based sensors. Springer, Berlin, Heidelberg, 155–176
17. Liedberg B, Nylander C, Lundström I (1995) Biosensing with surface plasmon resonance—how it all started. Biosens Bioelectron 10(8):i–ix
18. Lundström I (2014) From a laboratory exercise for students to a pioneering biosensing technology. Plasmonics 9(4):741–751
19. Hanson EK, Whelan RJ (2023) Application of the Nicoya OpenSPR to studies of biomolecular binding: a review of the literature from 2016 to 2022. Sensors (Basel) 23(10):4831

Chapter 18

Proximity Ligation Assay to Determine Protein–Protein Interactions in Muscle Stem Cells

Marie J Catenacci and Michael A. Rudnicki

Abstract

Satellite cells are present only in skeletal muscle and are crucial for regeneration. In recent years, satellite cell heterogeneity has identified different subpopulations, such as bona fide muscle stem cells. These populations are rare and in small numbers, making studying them very difficult. Moreover, satellite cells are sensitive to the environment and niche. To avoid niche disruption and changes to the cell state, our lab developed a mix of different techniques to study protein–protein interactions ex vivo on cultured myofibers. This protocol is based on the Proximity Ligation Assay (PLA) technique and uses IMARIS software to analyze protein interactions.

Keywords Skeletal muscle, Satellite cells, Niche, Protein complex, Protein interaction

1 Introduction

Skeletal muscle is composed of long multinucleated cells, termed myofibers and are closely associated with a small population of satellite cells, of which a subset represents true muscle stem cells. Satellite cells were first described by Mauro in 1961 and are only present in skeletal muscle tissue [1]. This population is crucial for growth and regeneration of skeletal muscle after a stress (i.e., injury) [2]. In the last 20 years, satellite cells have been widely described, and different subpopulations discovered [3–9]. These subpopulations showed unique properties (activation state, localization, stemness properties) but are often very rare and only present in small numbers in skeletal muscle. Muscle stem cells are located between the myofiber basal lamina and the sarcolemma, and typically very close to blood vessels. Indeed, this unique location makes the environment and the niche surrounding the cells major components influencing satellite cells shape as well as the intracellular pathways or protein interactions [2, 10, 11].

Joe Quadrilatero (ed.), *Myogenesis and Skeletal Muscle Biology*, Methods in Molecular Biology, vol. 3057,
https://doi.org/10.1007/978-1-0716-5396-8_18,

Several techniques used to study protein interactions in satellite cells need a large amount of material (i.e., Western Blot, co-immunoprecipitation) and have been done by dissociating satellite cells from their niche. More recently, the muscle stem cell field realized the importance of maintaining cells in their initial niche and had tried to develop techniques in this direction. However, low number of satellite cell subpopulations makes them very challenging to study in their environment.

In this context, to avoid niche disruption and study protein interactions in a small subpopulation, we developed a mix of different techniques. The current chapter describes methods to stain satellite cells and analyze protein interactions using Proximity Ligation Assay (PLA)

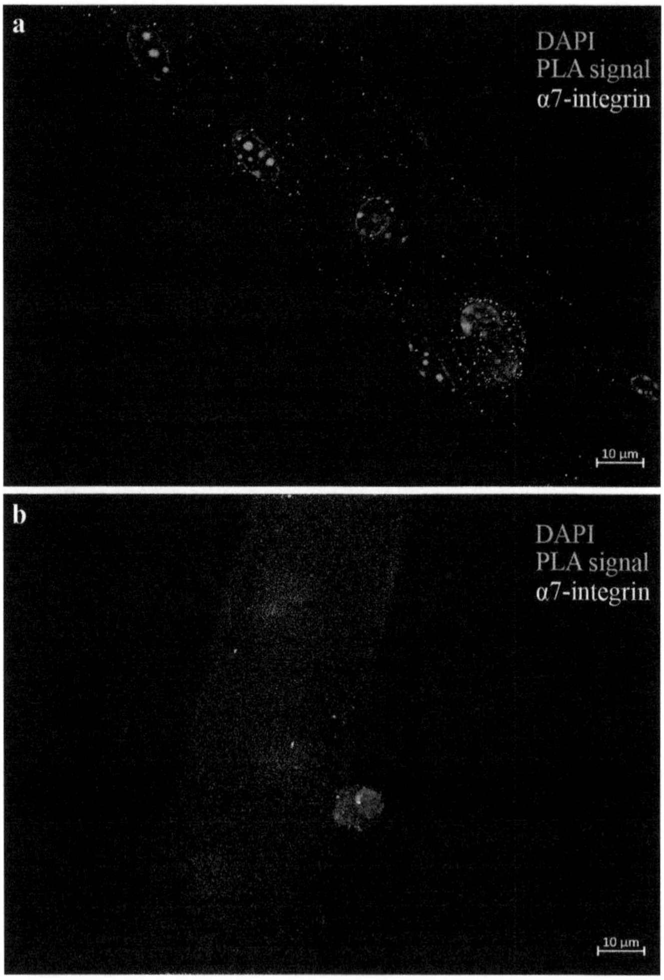

Fig. 1 Example of PLA between DMD and Mark2 in satellite cells on myofibers. (**a**) Image representative of PLA signal (orange) between DMD and Mark2 in WT fibers at 36 h of culture. (**b**) Image representative of PLA signal (orange) between DMD and Mark2 in *mdx* fibers at 36 h of culture. Scale bar: 10 µm

[12], on isolated EDL myofibers (Fig. 1). We have used this method extensively to quantitatively demonstrate diverse protein–protein interactions in situ in muscle stem cells within their niche [13–16]. This method is based on two probes and antibody incubation targeting two proteins of interest. Signal gets amplified when the two proteins are physically close to each other, and the probes can be linked. At the end of this chapter, we describe automatic analysis using IMARIS software.

2 Materials

Prepare all the solutions before starting the experiments, except the ligation and amplification solutions. All the reagents can be kept in the fridge.

2.1 Fiber Isolation, Fixation, and Transfer in IBIDI Plate

1. Quenching solution: 0.1 M glycine in 1× PBS (0.22 μm syringe filtered).
2. PBS 1×.
3. Pasteur pipette (*see* **Note 1**).
4. Fetal bovine serum (FBS): 0.22 μm syringe filtered.
5. 12-well plate IBIDI.

2.2 Blocking Permeabilization Solution

1. Blocking Permeabilization Solution (BPS): 2% goat serum, 1% bovine serum albumin, 50 mM glycine, 0.1% Triton X-100 in 1× PBS (0.22 μm syringe filtered) (*see* **Note 2**).

2.3 Antibody Staining and Probes Incubation

1. Antibody Staining Enhancer (ASE) 2×: 20 mM glycine, 0.1% Tween 20, 0.2% Triton and 0.2% H_2O_2 3% in PBS 1× (0.22 μ m syringe filtered) (*see* **Note 3**).
2. Primary antibodies (*see* **Notes 4 and 5**).
3. Probes: Probe anti-Mouse PLUS (Sigma-Aldrich) and Probe anti-Rabbit MINUS (Sigma-Aldrich) (see Table 1).
4. Duolink Antibody Diluent (1×) (Sigma-Aldrich).
5. 1.5 mL Eppendorf tubes.

Table 1
Preparation of probe solution

Probes	Host species	Product code	Company	Volume for 1 reaction (1 well)	Volume antibody diluent up to 60 μ L (1 well)
Probe anti-Mouse PLUS	Rb	DUO92001	Sigma	12 μL	
Probe anti-Rabbit MINUS	M	DUO92005	Sigma	12 μL	36 μL

2.4 Ligation and Amplification Step

1. Kit PLA (Sigma-Aldrich).
2. Wash buffer A and wash buffer B (Sigma-Aldrich) (*see* **Note 6**).
3. High-purity water.
4. 1.5 mL Eppendorf tubes.
5. Humidity chamber or incubator.
6. Cross-absorbed fluorescence-conjugated secondary antibodies (*see* **Note 7**).

2.5 DAPI Staining and Mounting Fibers

1. DAPI (4′, 6-diaminidino-phenylindole dihydrochloride).
2. Pasteur pipette.
3. Fetal bovine serum (FBS): 0.22 μm syringe filtered.
4. Glass coverslip (#1.5).
5. Hydrophobic PAP pen.
6. Superfrost microscope slides.
7. Clear nail varnish.
8. Aqueous mounting medium (*see* **Note 8**).

2.6 Satellite Cells PLA Imaging on Fibers Using Zeiss Zen Blue 2.6 Pro Software and Z1 Microscope

1. Epifluorescence microscope Z1.
2. Zeiss Zen Blue 2.6 Pro software.

2.7 Background Removal Using Deconvolution Tool in Zeiss Zen Blue 3.11 Software

1. Zeiss Zen Blue 3.11 software.

2.8 Analyzing Image with IMARIS Software

1. IMARIS software.

3 Methods

All the steps are performed while using the dissecting scope to avoid fiber aspiration.

3.1 Fiber Isolation, Fixation, and Transfer in IBIDI Plate

1. Fibers are isolated as mentioned in the protocol [17]. Briefly, after EDL muscle dissection and digestion in collagenase, fibers are released by flushing the EDL muscle. Fibers can be cultured in an incubator for different time points (as example 32 h, 48 h, 72 h). At the desired time point, fibers are fixed with 4% PFA.

2. After fixation, add Quenching solution to the well and rinse with PBS. The fibers can be kept in the fridge during days/months.
3. Coat 12-well plate IBIDI and Pasteur pipette with FBS for 2–5 min (*see* **Note 9**).
4. Aspirate FBS and transfer at least ten fibers per condition in the well using Pasteur pipette.

3.2 Blocking Permeabilization Solution

1. Cover the well with BPS solution and incubate for 2 h at room temperature on rocker.

3.3 Antibody Staining and Probes Incubation

1. Make 1× ASE.
2. Dilute primary antibodies in ASE.
3. Aspirate blocking buffer without touching fibers and add primary antibody solution without washing. Incubate the plate overnight at 4°C.
4. Wash two times with wash buffer A for 5 min.
5. Prepare probe solution (see Table 1).
6. Incubate fibers with probe solution (60 μL per well) during 1 h at 37°C in an incubator or in humidity chamber.

3.4 Ligation and Amplification Step

1. Wash two times with wash buffer A for 5 min.
2. During the wash, prepare ligation solution (see Table 2).
3. Add 60 μL of ligation buffer in each well and incubate 30 min at 37° in incubator or in humidity chamber.
4. Wash two times with Wash Buffer A for 5 min.
5. During the wash, prepare amplification solution (see Table 2).
6. Add secondary antibody to the polymerase solution (*see* **Note 10**).

Table 2
Preparation of ligation and amplification solution

	Reagent	Volume for 1 reaction (1 well)
Ligation solution	Ligation buffer 5×	12 μL
	Sterile water	46.5 μL
	Ligase	1.5 μL
Amplification solution	Polymerase buffer 5×	12 μL
	Sterile water	47.25 μL
	Polymerase	0.75 μL

7. Add 60 µL of polymerase buffer and incubate 1 h at 37°C in an incubator or in a humidity chamber, protected from the light (*see* **Note 11**).
8. Wash two times with wash buffer B for 10 min.
9. Dilute wash buffer B 1/100 in high-purity water.
10. Wash two times with diluted 1/100 wash buffer B for 5 min.

3.5 DAPI Staining and Mounting Fibers

1. Prepare DAPI solution.
2. Add DAPI solution and incubate 5 min at room temperature, in the dark.
3. Surround the slide with a hydrophobic barrier using a PAP pen.
4. Aspirate DAPI solution along with fibers using FBS-coated Pasteur pipette.
5. Transfer fibers in DAPI solution on a slide.
6. Aspirate liquid as much as possible.
7. Place one to two droplets of mounting media (*see* **Note 8**). Carefully place a coverslip on the slide. Avoid introducing air bubbles on fibers. Let the slides dry in the dark for 1–2 h. Slides should be stored at 4°C and visualized within 1–2 weeks.

3.6 Satellite Cells PLA Imaging on Fibers Using Zeiss Zen Blue 2.6 Pro Software and Z1 Microscope

1. Place the slide, using 63× oil immersion lens, do not forget to add a drop of immersion oil (*see* **Note 12**).
2. Use "locate" tab (white arrow in Fig. 2) to localize fiber using DAPI channel.
3. Switch to "Acquisition" tab (red arrow in Fig. 2) and press Live (green arrow in Fig. 2).
4. Locate α7-integrin$^+$ satellite cells and center them in the field of view.
5. Adjust Image settings to optimize laser intensity and time exposure for each channel (blue arrows in Fig. 2) (*see* **Note 13**).
6. To capture PLA signal in whole satellite cell, use "Z-Stack" tool to capture image slices from the upper to the lower limits of the cells using "First/Last" box (red box in Fig. 2).
7. Image Z-stack with all three channels turned on (DAPI, α7-integrin$^+$, PLA) (white box in Fig. 2) by pressing "Start Experiment" button in (yellow arrow in Fig. 2).
8. Save image in .czi format.

Fig. 2 Z1 microscope and z-stack imaging of PLA in satellite cells. Screenshot view of Zen software control panel with parameters used to image PLA in satellite cells. White arrow: "Locate" tab to find the fiber using DAPI channel. Red arrow: "Acquisition" tab. Green arrow: "Live" Button allowing live viewing. Blue arrow: "Channels" parameters used to modify channel intensity and exposure time for each individual channel. Red Boxes: "Z-Stack" window used to set the parameters for the upper and lower limits of cell of interest. Yellow arrow: "Start Experiment" Button used to capture the Z-stack image once all the parameters have been selected

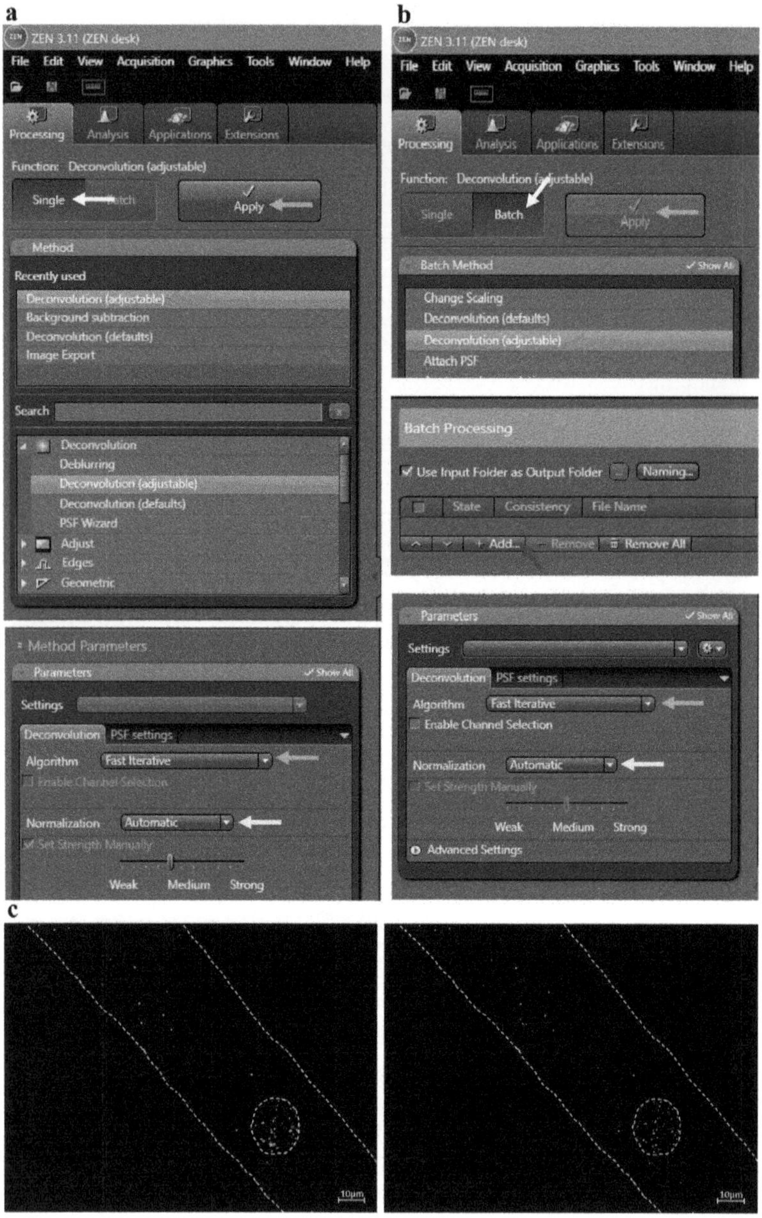

Fig. 3 Deconvolution using Zen software. Screenshot view of Zen software control panel with parameters used to remove the background. (**a**) Option 1: "single" tool to apply the parameters on one picture. (**b**) Option 2: "batch" tool to apply the parameters on selected pictures. Arrows in each option correspond to different steps, explained in the protocol. (**c**) Image before (left panel) and after (right panel) deconvolution protocol. PLA was performed between Mark2 and DMD proteins. White dashed line represents the fiber and white dashed circle represents satellite cell

3.7 Background Removal Using Deconvolution Tool in Zeiss Zen Blue 3.11 Software

Two choices are possible: Option 1 to perform the deconvolution on one picture ("Single") and Option 2 to perform it on a batch of pictures ("Batch").

Option 1:

a. Open the pictures in Zeiss Zen Blue 3.11 software.
b. Click on "Single" tool (white arrow in Fig. 3a) and select "Deconvolution adjustable" option (red arrow in Fig. 3a) in the "Processing" tab.
c. Select "Fast Iterative" for the algorithm option (green arrow in Fig. 3a) and choose automatic for the normalization (yellow arrow in Fig. 3a) in "Method Parameters" box (*see* **Note 14**).
d. Press Apply (blue arrow in Fig. 3a).
e. Save the image in .czi format.

Option 2:

a. Click on "Batch" tool (white arrow in Fig. 3b) and select "Deconvolution adjustable" option (red arrow in Fig. 3b) in the "Batch Method" tab.
b. Add all the pictures for the analysis by pressing on "+" in "Batch Processing" box (purple arrow in Fig. 3b).
c. Select the same settings as Option1 in "Method Parameters" box (green and yellow arrows in Fig. 3b). Click right, select "Copy parameters" and paste the parameters in all pictures.
d. Press Apply (orange arrow in Fig. 3b) (*see* **Note 15**).

3.8 Analyzing Image with IMARIS Software

Nonautomatic analysis can be realized by using ZEN software and "Graphics" tool. This part describes automatic analysis using IMARIS software.

1. Open the image in IMARIS software.
2. In "Arena" tab, select the folder of interest (white arrows in Fig. 4a).
3. Double click on the picture to switch in "Surpass" tab (red arrows in Fig. 4b).
4. Click on "add new cells" (red arrows in Fig. 4b) to start analyzing. To count the number of dots, follow these next ten steps.
5. Step 1: In the favorite parameters, choose "default" and select the first box "Detect Nucleus, Cell and Vesicle" in "select detection type" tool (green arrow in Fig. 4c) (*see* **Note 16**).
6. Step2: In the "Nuclei detection" section, choose DAPI as "source channel" and add nucleus diameter (*see* **Note 17**) (Fig. 4d).

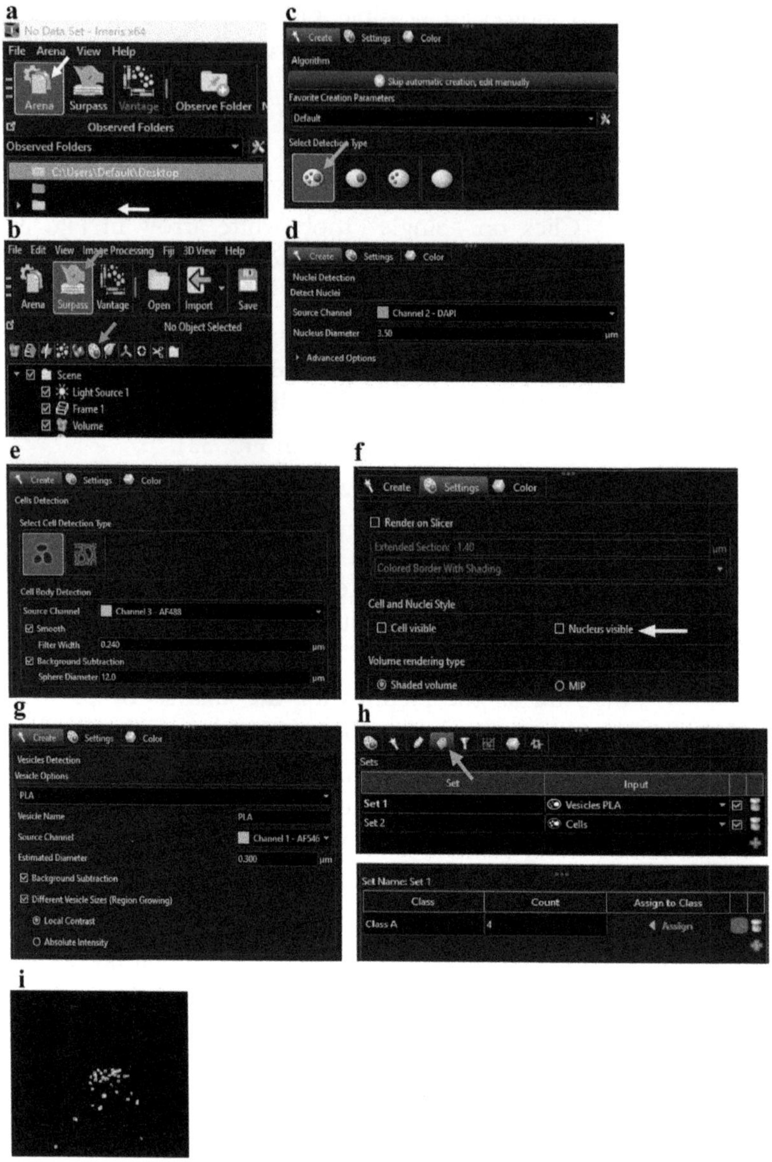

Fig. 4 Analysis of PLA signal in IMARIS software. (**a**) White arrow: "Arena" tab to localize the folder of interest. (**b**) Red arrows: in "surpass" tab, press "add new cells" tool to start the analysis. (**c**) Green arrow: select correct setting to start the analysis. (**d**) This step allows to detect nuclei. (**e**) This step allows to detect satellite cells. (**f**) Yellow arrow: unselect these two conditions to see PLA dots. (**g**) This step allows to detect vesicles in the cells selection. Click on the green double arrow to end the analysis. (**h**) Pink arrow: "tag" allows to select only the dots inside the satellite cell. (**i**) Image view on PLA between Mark2 and DMD after analysis using IMARIS software

7. Step3: Change threshold to select all nuclei (*see* **Note 18**).
8. Step 4: In "filter Nuclei" section, choose only the nuclei of satellite cell (*see* **Note 19**).
9. Step 5: For cell detection, select the first box and add the source channel (AF488, corresponding to alpha7-integrin channel). Select "smooth" and "background subtraction" (*see* **Note 20**) (Fig. 4e).
10. Step 6–7: As nuclei detection process, correct the threshold and choose satellite cell.
11. In "setting" tab, unclick "cell visible" and "nucleus visible" to see the dots in AF546 (yellow arrow in Fig. 4f).
12. Step 8: in "vesicle detection" section, choose the source channel (AF546) and estimated diameter (*see* **Note 21**) (Fig. 4g).
13. Step 9: To filter vesicles, select the option "Distance from origin" and adjust the selection to choose all the vesicles in the satellite cells and remove the others (on fibers, for example).
14. Step 10: Adjust threshold and press the double green arrow at bottom of the box to finalize the analysis.
15. If some vesicles are selected and appear outside the satellite cells, you can select only the vesicles inside the cells using "Tag tab" (pink arrow in Fig. 4h). Add a set and select "vesicles PLA". Select the vesicles on the image in the cells and in the following table (named Set Name), press "assign". The number corresponds to vesicles number inside the satellite cells (*see* **Note 22**).
16. Analysis can be found in Properties tab and can be exported.

4 Notes

1. To transfer fibers, use Pasteur pipette previously curve under a flame. Add cotton at the opening (top of the Pasteur pipette) to avoid contamination.
2. When Blocking Permeabilization Solution is stored longer than 1 week, either add 0.02% sodium azide or freeze aliquots at −20°C.
3. Dilute 2× ASE in PBS 1× or directly in hybridoma antibody (DSHB).
4. When planning primary antibody panel, ensure the primary antibodies are raised in two different species (i.e., Mouse and Rabbit). The probes do not exist with different IgG isotypes (i.e., IgG, IgG2a, IgG2b…).

5. Primary antibodies need to be validated in regular immunostaining. For PLA experiment, different dilutions need to be tested for each antibody to find the proper combination. Usually, the antibody dilution is lower for PLA compared to immunostaining to avoid background as much as possible. Differences between batches can be observed and validation needs to be done on every batch by regular immunostaining. A cell lineage marker should be used to identify specific cell types (as satellite cells with α7-integrin).
6. Possibility to do homemade wash buffer (wash buffer A: 10 nM Tris pH = 7.4, 150 mM NaCl, and 0.05% Tween/ wash buffer B: 200 mM Tris pH = 7.5, 100 mM NaCl).
7. Generally, our lab uses Alexa Fluor dyes, as most are bright and photostable, and the Fisher Scientific website provides information on stability in buffer and fluorophore brightness to dye recommendation based on your filter set and more. The PLA kit is read in the 546 channels. The most common panel our lab uses for PLA is Alexa Fluor 488 coupled to channel 546, used by PLA.
8. PermaFluor (Fisher Scientific, TA-030-FM) is our preferred mounting media, but other mounting media can be substituted.
9. The coating with FBS allows fibers to remain in suspension and avoid them to stick to the well.
10. During the polymerase step, add secondary antibody to localize the interested subpopulation on the fiber. In this example, our lab used α7-integrin to stain the satellite cells. For the satellite cells markers, do not use same species as those used for the PLA (For the example experiment, no mouse or rabbit).
11. Polymerase incubation time may vary from interaction to interaction. The background of PLA can depend on this step. If the background is too high, you can decrease incubation time. For some interactions, it is necessary to increase this time up to 90–120 min.
12. The 63× oil immersion lens allows for a resolution necessary for the analysis. A 40× oil immersion can be substituted.
13. In our lab, image settings were adjusted and optimized using the Zeiss Zen Blue 2.6 Pro software.
14. For the normalization, our lab selects the automatic option around 4.
15. In Batch option, all pictures are automatically saved in the same folder.
16. Between each step, click on the blue arrow (at the bottom of the box). If you click on the green arrow, all the steps are done automatically, and the parameter cannot be changed at each step.

17. Our lab chooses a diameter between 3.5 and 5.00 μm for satellite cell nuclei.
18. All the nuclei (satellite cells and fibers) are delineated.
19. If the filter type does not appear, click on "add" and select the nucleus of interest.
20. In our example, we put 0.240 μm for "Smooth" and 12.0 μm for the "background subtraction".
21. Our lab chooses 0.300 μm.
22. With this process, you can assign the dots inside the cell or separate the dots between cytoplasm and nucleus by pressing "plus" in the "Set Name" table to increase the number of categories.

References

1. Mauro A (1961) Satellite cell of skeletal muscle fibers. J Biophys Biochem Cytol 9:493–495
2. Yin H, Price F, Rudnicki MA (2013) Satellite cells and the muscle stem cell niche. Physiol Rev 93(1):23–67. https://doi.org/10.1152/physrev.00043.2011
3. Chakkalakal JV et al (2014) Early forming label-retaining muscle stem cells require p27kip1 for maintenance of the primitive state. Development 141(8):1649–1659. https://doi.org/10.1242/dev.100842
4. Kuang S, Kuroda K, Le Grand F, Rudnicki MA (2007) Asymmetric self-renewal and commitment of satellite stem cells in muscle. Cell 129(5):999–1010. https://doi.org/10.1016/j.cell.2007.03.044
5. García-Prat L et al (2020) FoxO maintains a genuine muscle stem-cell quiescent state until geriatric age. Nat Cell Biol 22(11):1307–1318. https://doi.org/10.1038/s41556-020-00593-7
6. Rocheteau P et al (2012) A subpopulation of adult skeletal muscle stem cells retains all template DNA strands after cell division. Cell 148(1-2):112–125. https://doi.org/10.1016/j.cell.2011.11.049
7. Scaramozza A et al (2019) Lineage tracing reveals a subset of reserve muscle stem cells capable of clonal expansion under stress. Cell Stem Cell 24(6):944–957.e945. https://doi.org/10.1016/j.stem.2019.03.020
8. Vartanian AD et al (2019) PAX3 confers functional heterogeneity in skeletal muscle stem cell responses to environmental stress. Cell Stem Cell 24(6):958–973.e9. https://doi.org/10.1016/j.stem.2019.03.019
9. Guilhot C, Catenacci M, Lofaro S, Rudnicki MA (2024) The satellite cell in skeletal muscle: a story of heterogeneity. Curr Top Dev Biol 158:15–51. https://doi.org/10.1016/bs.ctdb.2024.01.018
10. Feige P, Brun CE, Ritso M, Rudnicki MA (2018) Orienting muscle stem cells for regeneration in homeostasis, aging, and disease. Cell Stem Cell 23(5):653–664. https://doi.org/10.1016/j.stem.2018.10.006
11. Feige P, Rudnicki MA (2018) Muscle stem cells. Curr Biol 28(10):R589–R590. https://doi.org/10.1016/j.cub.2018.02.064
12. Alam MS (2018) Proximity ligation assay (PLA). Curr Protoc Immunol 123(1):e58. https://doi.org/10.1002/cpim.58
13. Dumont NA et al (2015) Dystrophin expression in muscle stem cells regulates their polarity and asymmetric division. Nat Med 21(12):1455–1463. https://doi.org/10.1038/nm.3990
14. Chang NC et al (2018) The dystrophin glycoprotein complex regulates the epigenetic activation of muscle stem cell commitment. Cell Stem Cell 22(5):755–768.e756. https://doi.org/10.1016/j.stem.2018.03.022
15. Wang YX et al (2019) EGFR-Aurka signaling rescues polarity and regeneration defects in dystrophin-deficient muscle stem cells by increasing asymmetric divisions. Cell Stem Cell 24(3):419–432.e416. https://doi.org/10.1016/j.stem.2019.01.002
16. Esper ME et al (2025) Intrinsic muscle stem cell dysfunction contributes to impaired regeneration in the mdx mouse. J Cachexia Sarcopenia Muscle 16(1):e13682. https://doi.org/10.1002/jcsm.13682
17. Pasut A, Jones AE, Rudnicki MA (2013) Isolation and culture of individual myofibers and their satellite cells from adult skeletal muscle. J Vis Exp 73:e50074. https://doi.org/10.3791/50074

GPSR Compliance
The European Union's (EU) General Product Safety Regulation (GPSR) is a set of rules that requires consumer products to be safe and our obligations to ensure this.

If you have any concerns about our products, you can contact us on

ProductSafety@springernature.com

In case Publisher is established outside the EU, the EU authorized representative is:

Springer Nature Customer Service Center GmbH
Europaplatz 3
69115 Heidelberg, Germany

www.ingramcontent.com/pod-product-compliance
Ingram Content Group UK Ltd.
Pitfield, Milton Keynes, MK11 3LW, UK
UKHW050613241026
476948UK00005B/176
9781071653951